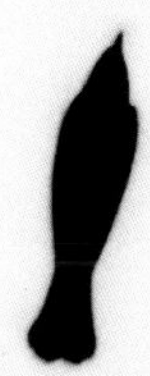

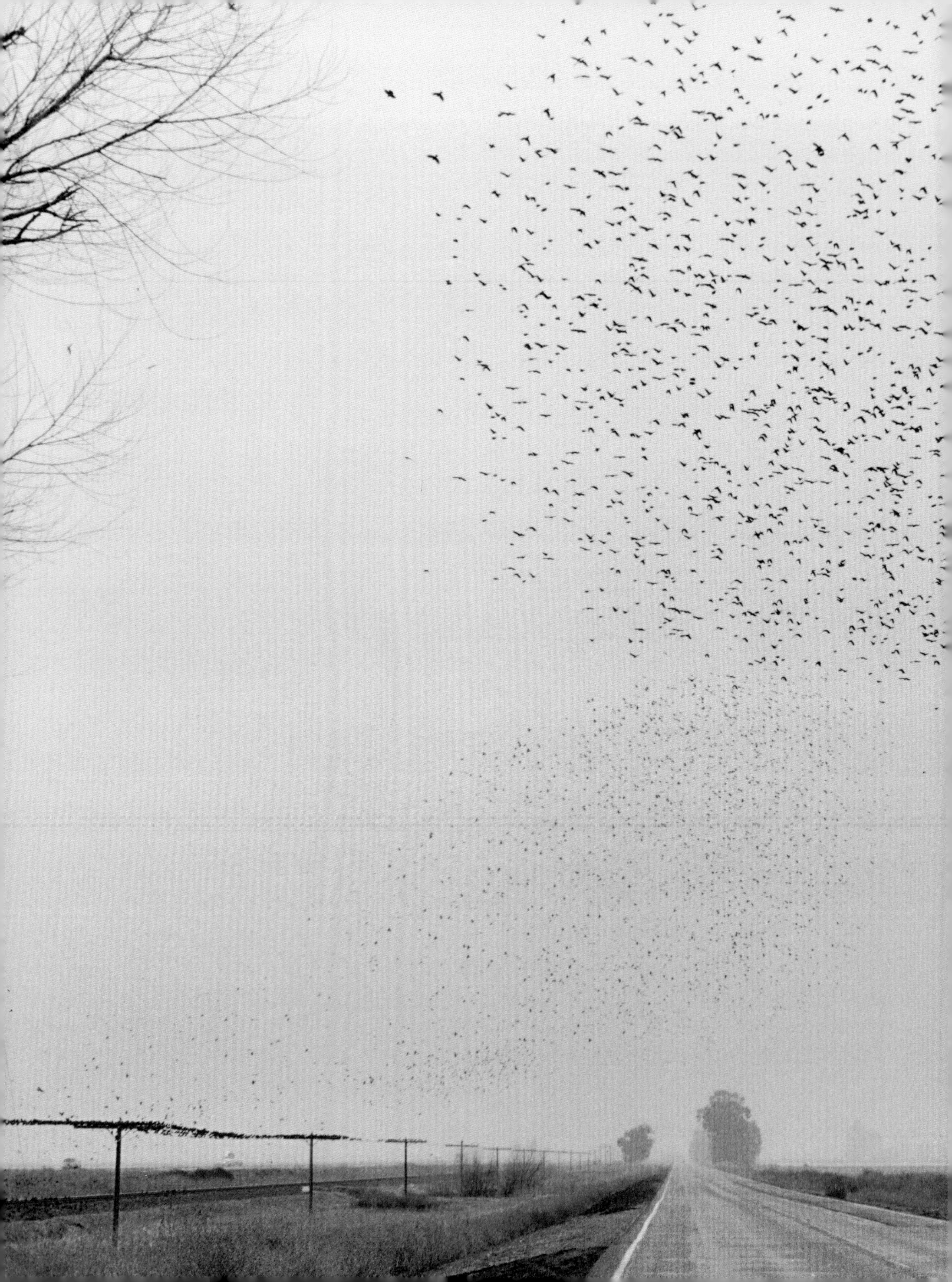

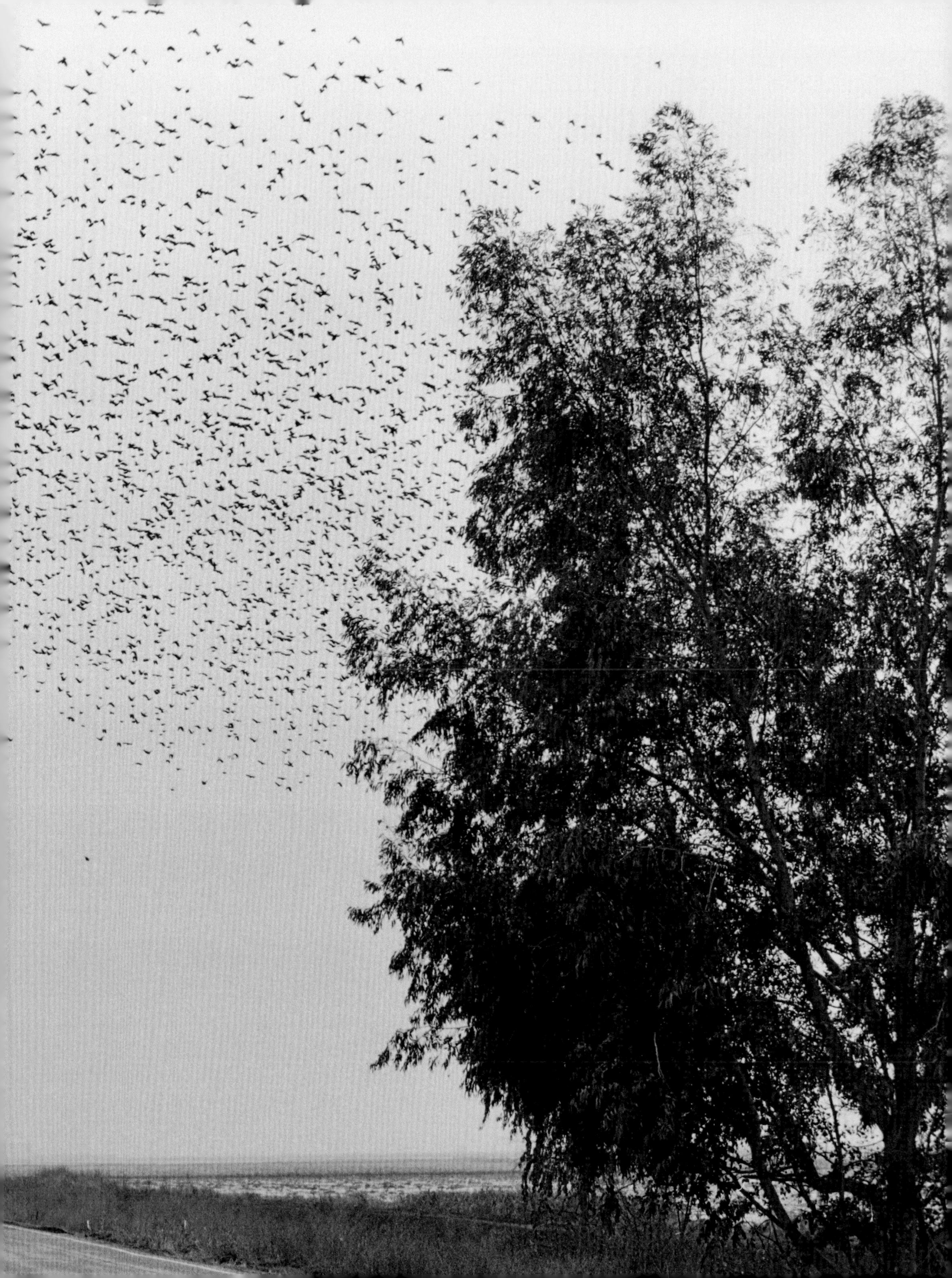

Lukas F

elzmann

Swarm

With texts by:
Peter Pfrunder
Gordon H. Orians
Deborah M. Gordon
Wallace Stevens
Lukas Felzmann

Lars Müller Publishers

Does Nature Play?

Peter Pfrunder

Various are the roads of man. He who follows and compares them will see strange figures emerge, figures which seem to belong to that great cipher which we discern written everywhere, in wings, eggshells, clouds and snow, in crystals and in stone formations, on ice-covered waters, on the inside and outside of mountains, of plants, beasts and men, in the lights of heaven, on scored disks of pitch or glass or in iron filings round a magnet, and in strange conjectures of chance. In them we suspect a key to the magic writing, even a grammar, but our surmise takes on no definite forms and seems unwilling to become a higher key.

Novalis, *The Novices of Sais*, 1802

We hear them flapping, we hear them screeching. Sometimes quieter and softer, sometimes more intense and more shrill, then the clamor ebbs away into a distant, background noise—and then rises again suddenly. A senseless tangle of agitated birdsong, a deafening din that one cannot escape.

It might sound like a paradox, but viewing the *Swarm* photographs of Lukas Felzmann is also an *acoustic* experience. Even if the pictures seem poor in content at first glance—just birds, sky, earth—they acquire a richness through their dense sequencing, the rhythm of near and far, the alternation of profusion and sparseness, the dynamic of small and large changes, and through the variety of that which is always the same. Isn't *Swarm* an exceptionally musical book? A score in which the visual experience almost cries out for an acoustic translation? I wouldn't be surprised if I came across this exceptional photo series again one day in musical form—not an authentic recording of birdsong, but as an instrumental performance based on the photographic original. The Blackbird Symphony.

The buzzword is "swarm intelligence," which has acquired an unforeseen reality in the era of Facebook and Twitter. The behavior of a collective without a center has become a social phenomenon, which is of interest not only to natural scientists, but also in particular to politicians and economists. Why does a mass of individuals turn in one direction or the other? Who or what tips the balance? How much room for maneuver, how much freedom do individuals have over their decisions? What has to happen for a new formation to form out of another? Why do groups form? And why do they disperse again? As always, when we observe great historical and social revolutions, when new phenomena emerge, we want to understand. To find explanations for the unknown. To get to grips with the mysterious, by giving it a name at least.

Who knows, maybe a scientific analysis of Lukas Felzmann's images would shed some light on the rules that flocks of birds use when they swarm through the air, drop and climb again, gather together, and drift apart. One could measure the density of the flocks, determine the direction of their flight paths, plot the different patterns of the formations. But that's not what this is about. When we leaf through from page to page, inquisitively follow the astonishing movements of the birds, and let ourselves be driven through the air with them, we are looking for something different. Something that has more to do with poetry than scientific observation.

It's as if we are eavesdropping on a foreign language. A language that consists of signs and sounds that have never been heard before. We attempt to understand messages whose meaning we are unable to grasp. We infer a system, assume a structure. But where does order begin? And where does the chaos end? Can we recognize movements that have any kind of purpose? Or are we witnesses of an aimless, natural game? Does nature play?

Perhaps the reason why the *Swarm* series is so fascinating is precisely because it depicts something uncontrollable, irrational, free in our thoroughly rationalized world. Because the signatures that the birds write in the sky don't allow any kind of clear interpretation. This is where the work of the photographer comes into play: with his camera he reveals things that the human eye perceives at best fleetingly and unconsciously. He collects signs, bestowed on him by chance—mysterious hieroglyphics for an imaginary archive. And he invites us to read these symbols as aesthetic figures, as a language of the unconscious. Or even as a magical, strange music. Fleeting and recurring, sweet and threatening, simple and complex all at the same time. And without any sort of purpose.

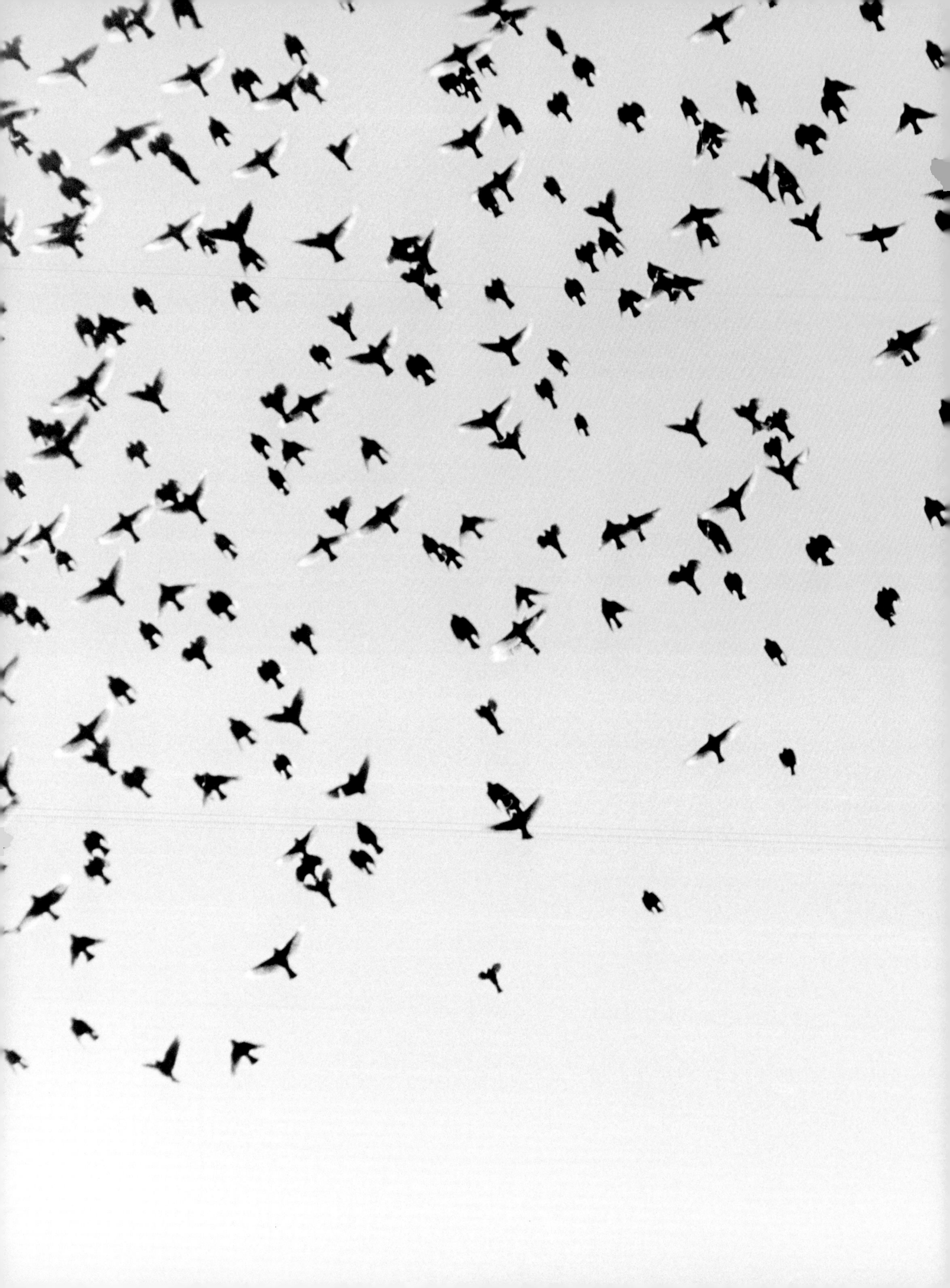

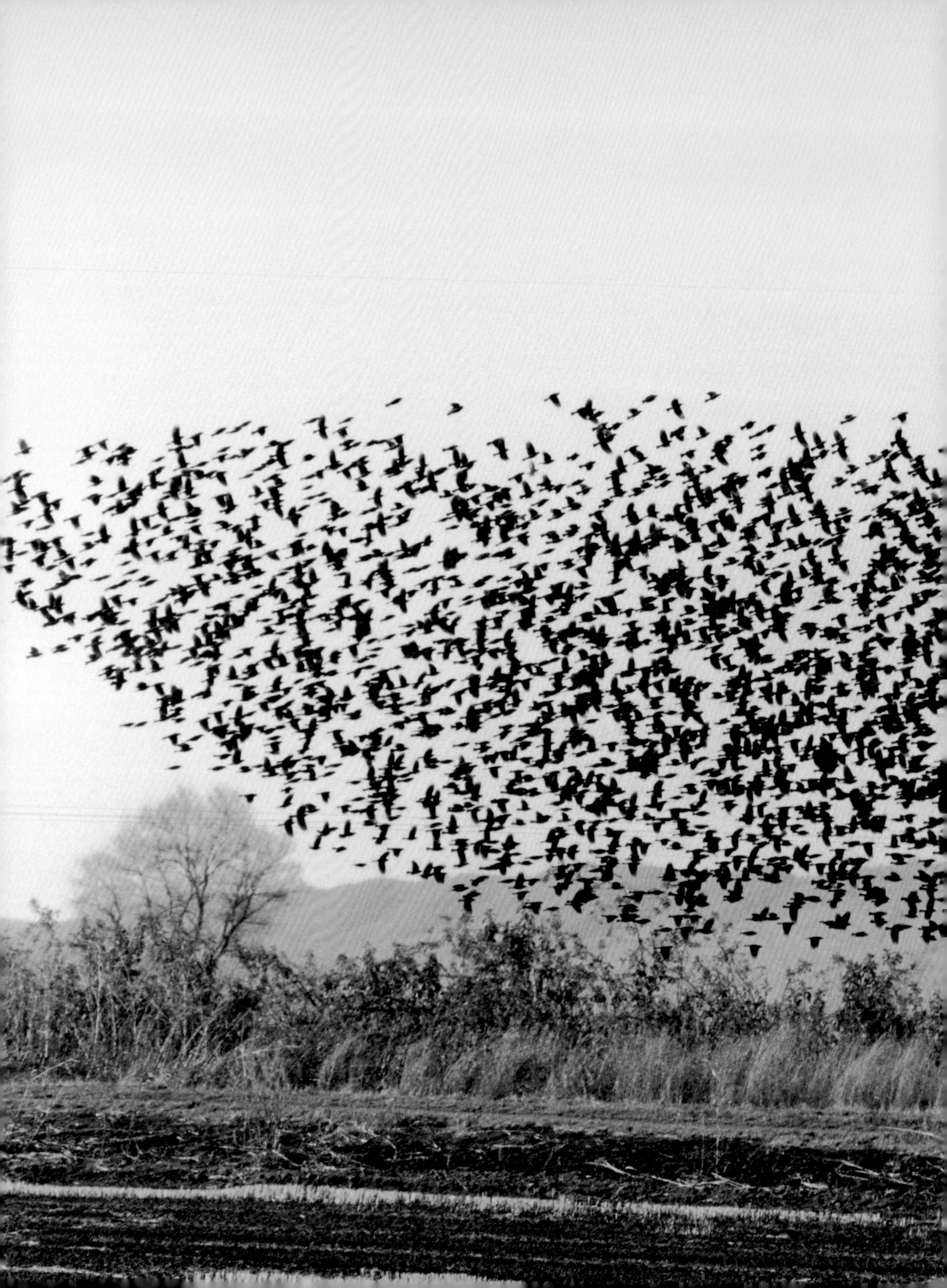

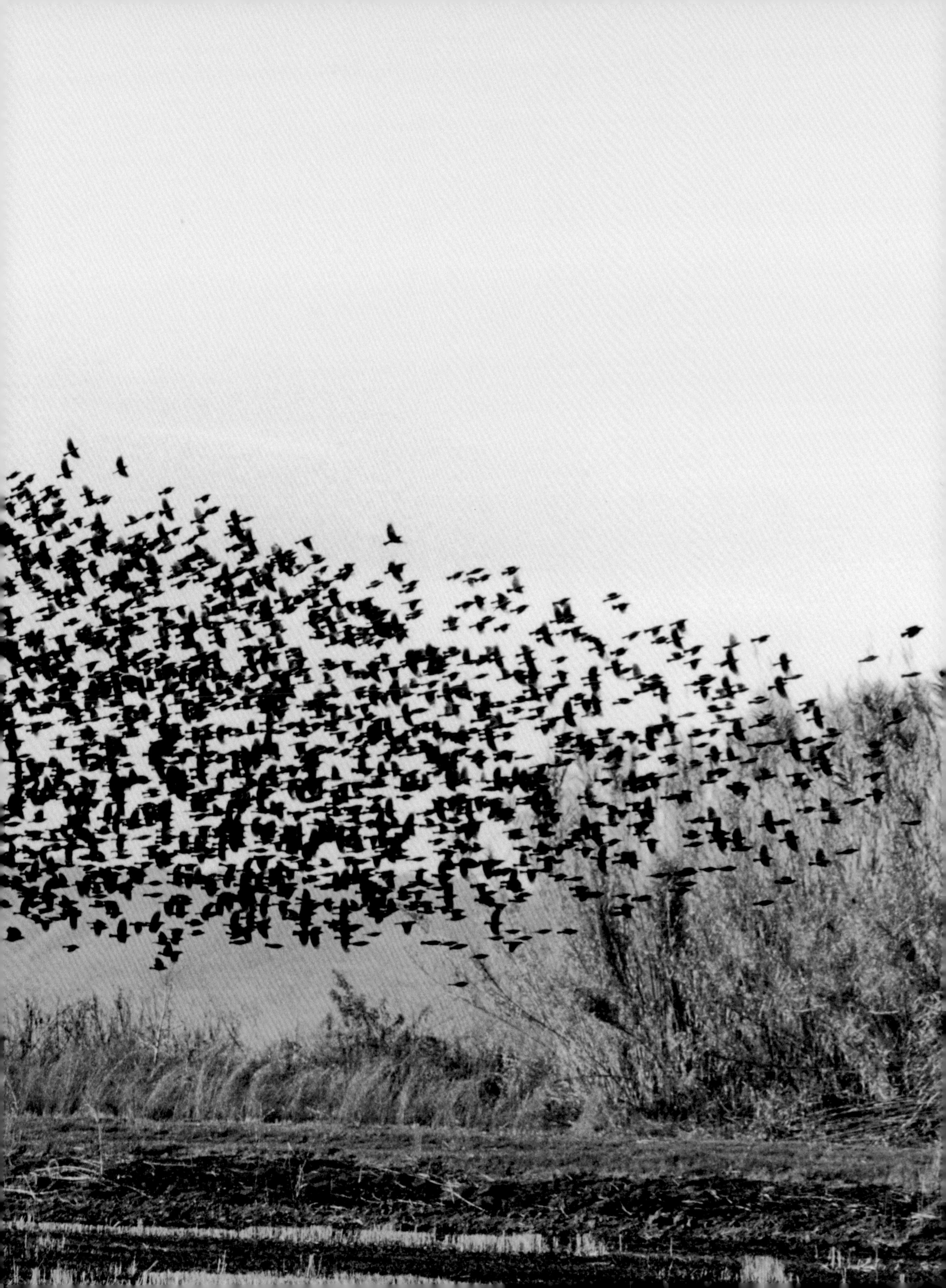

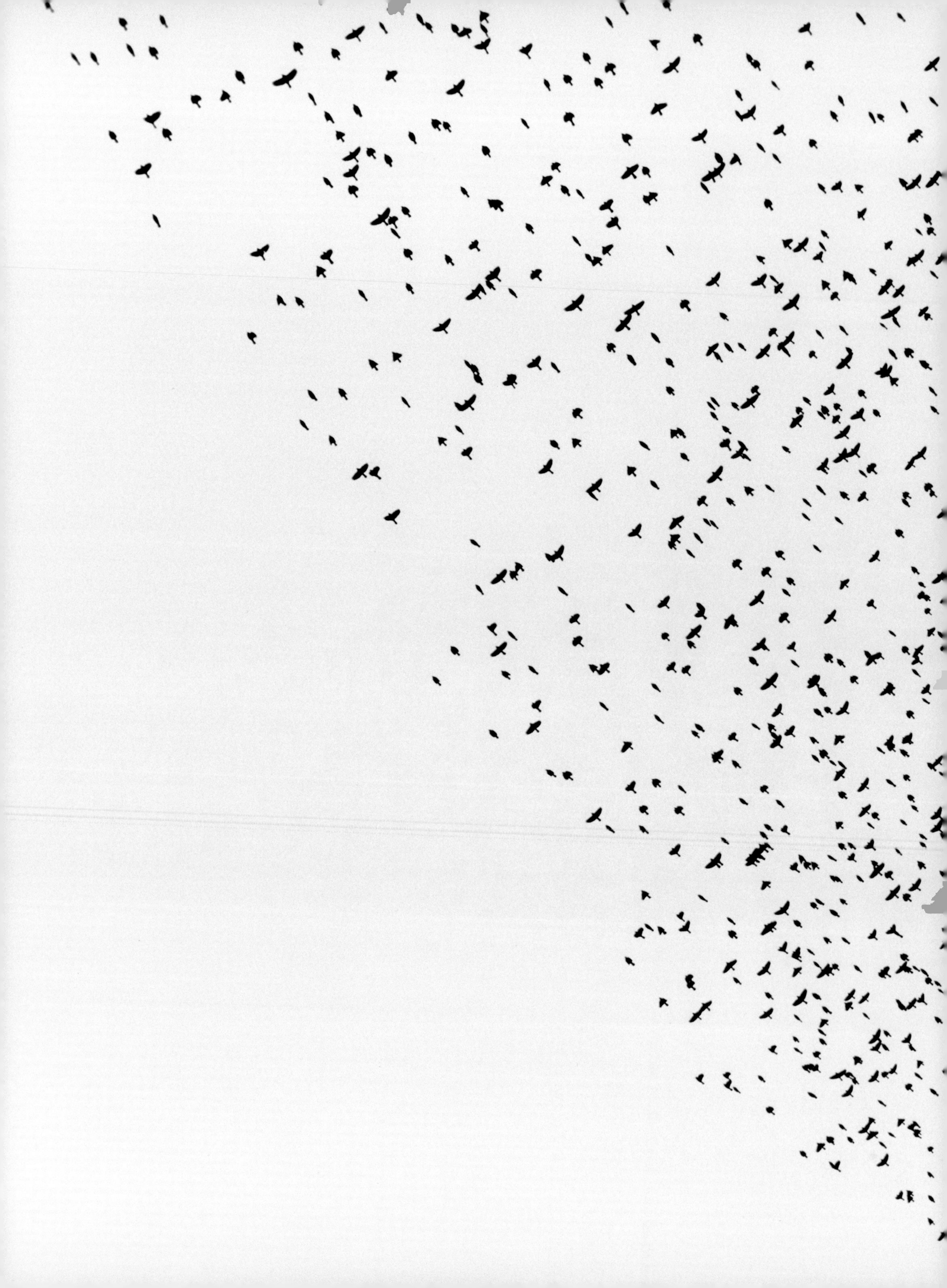

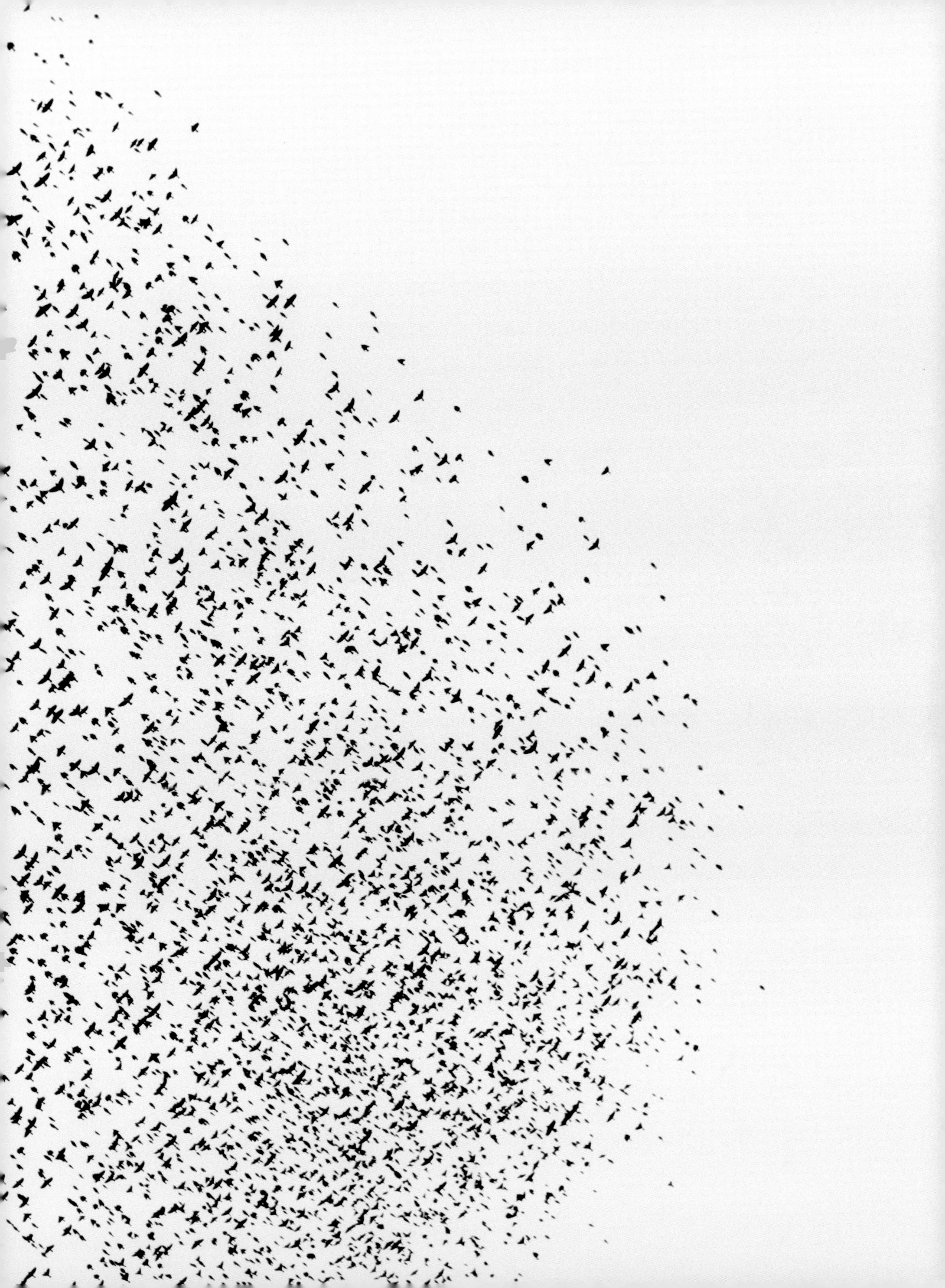

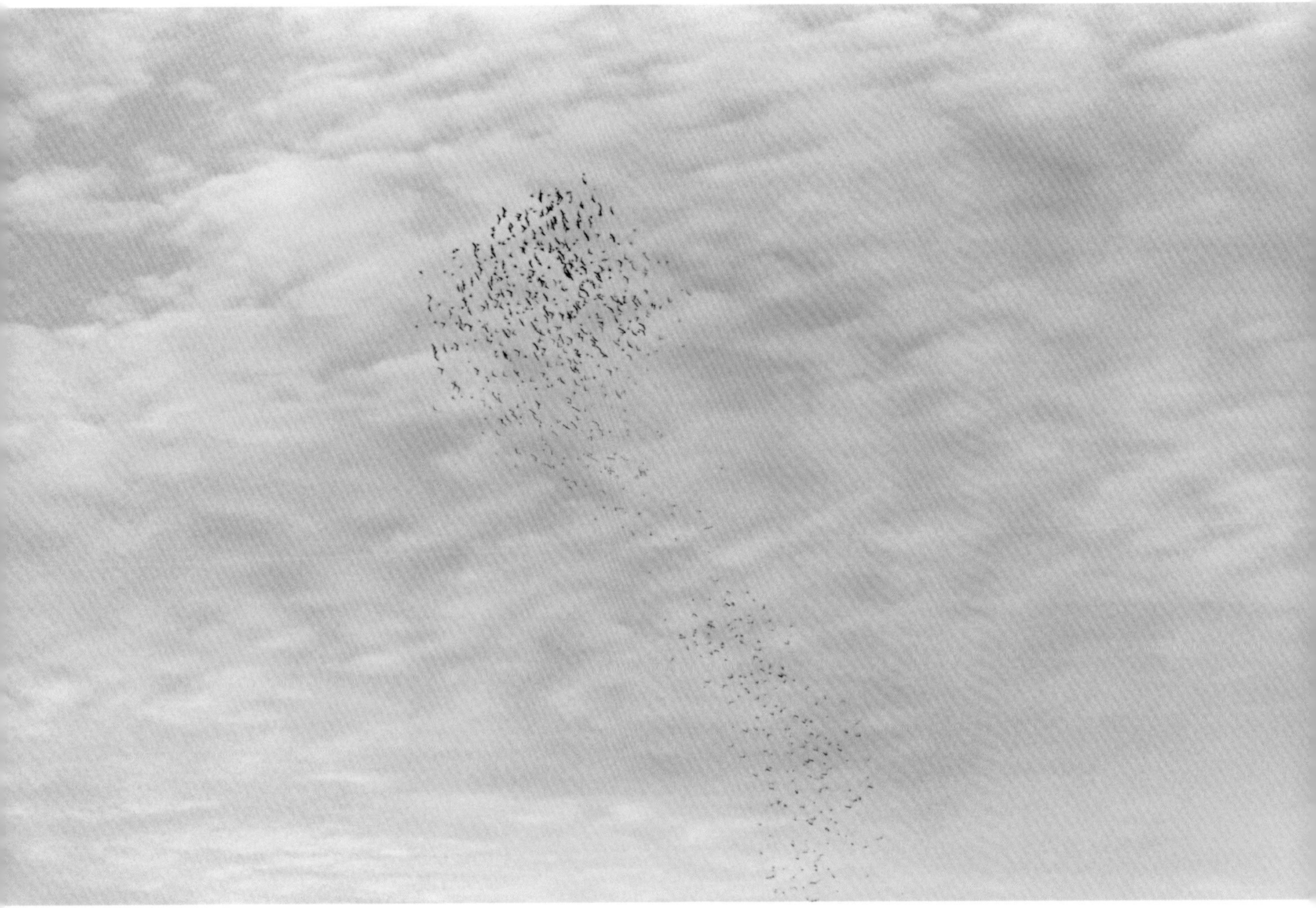

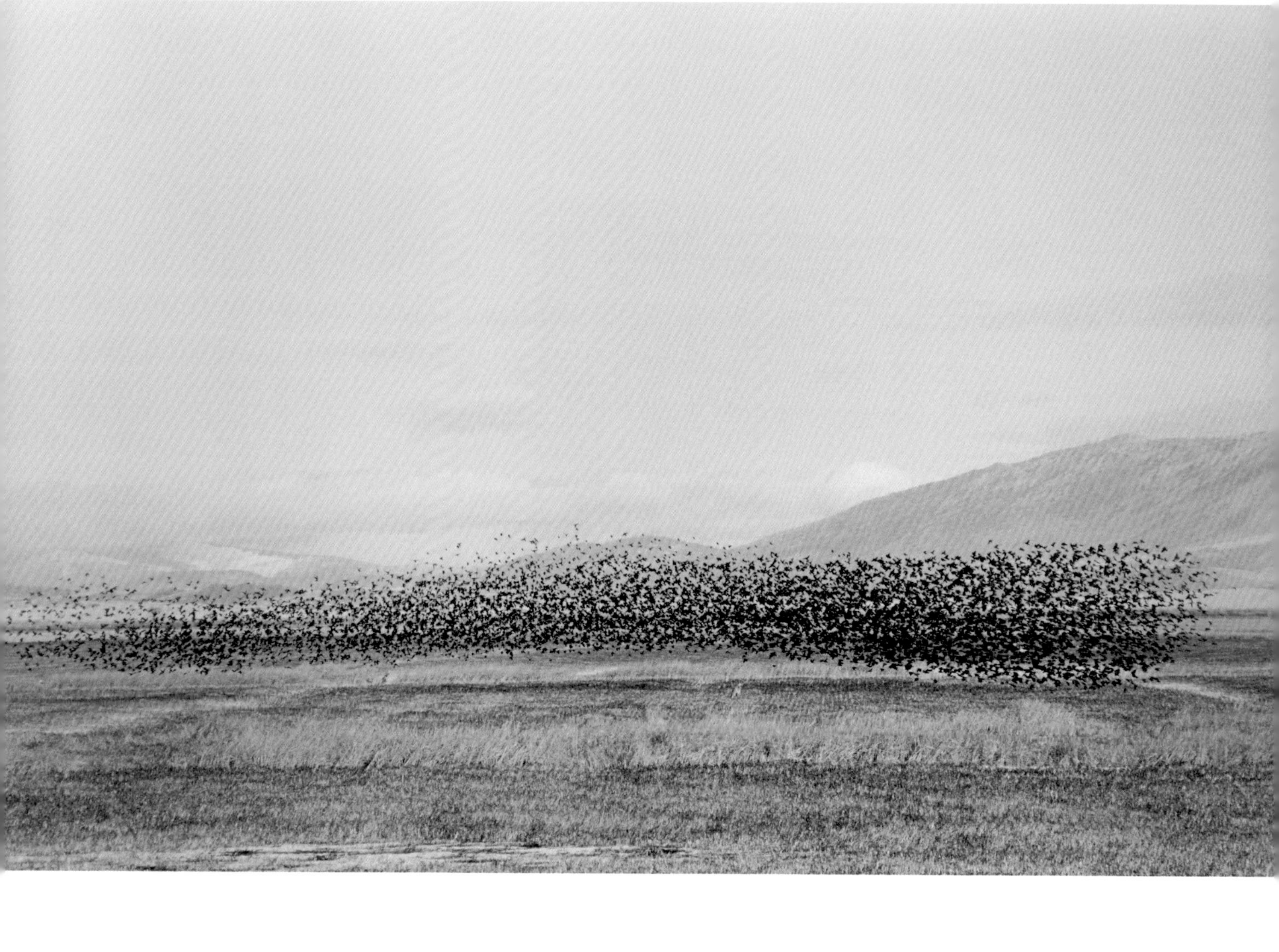

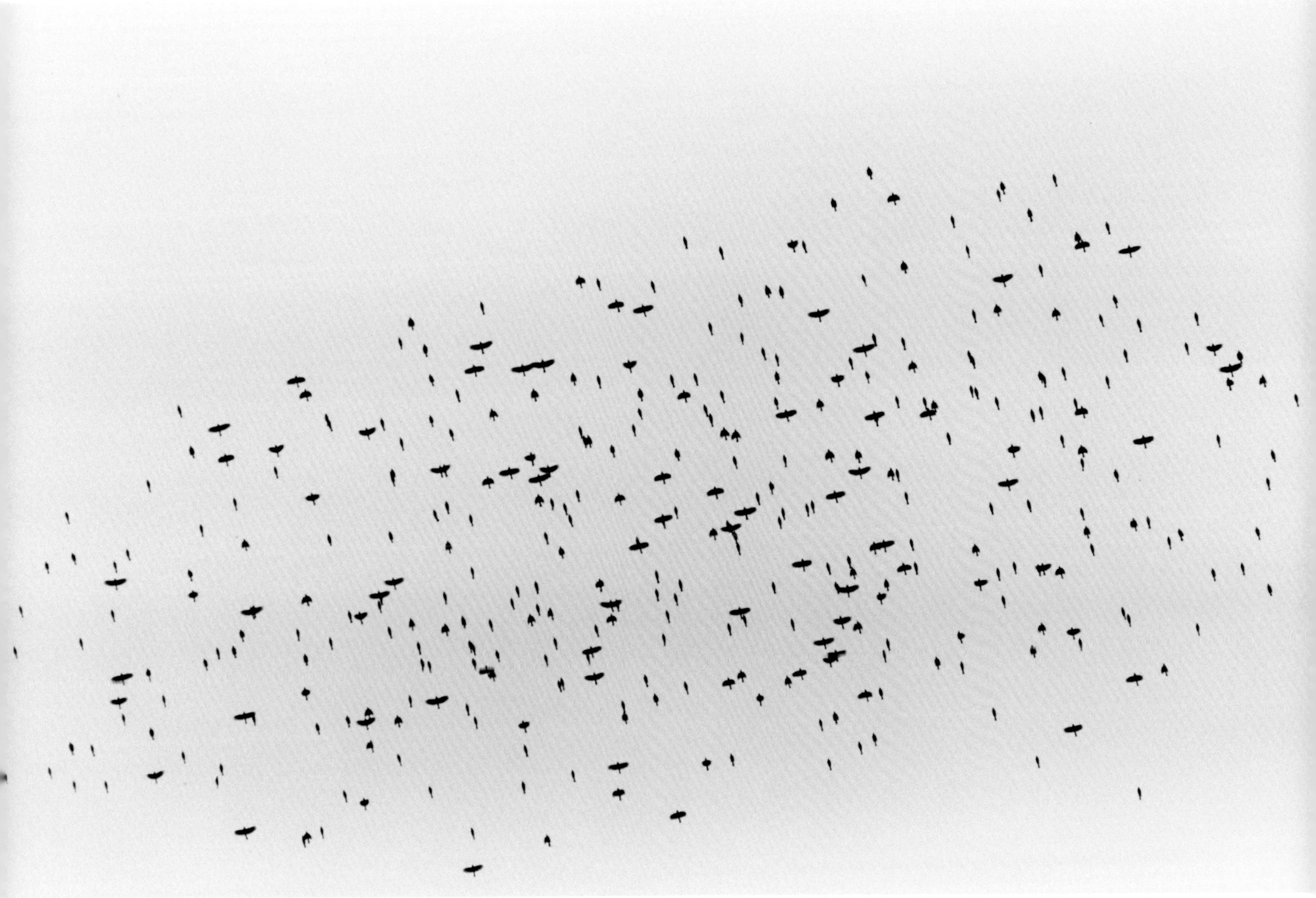

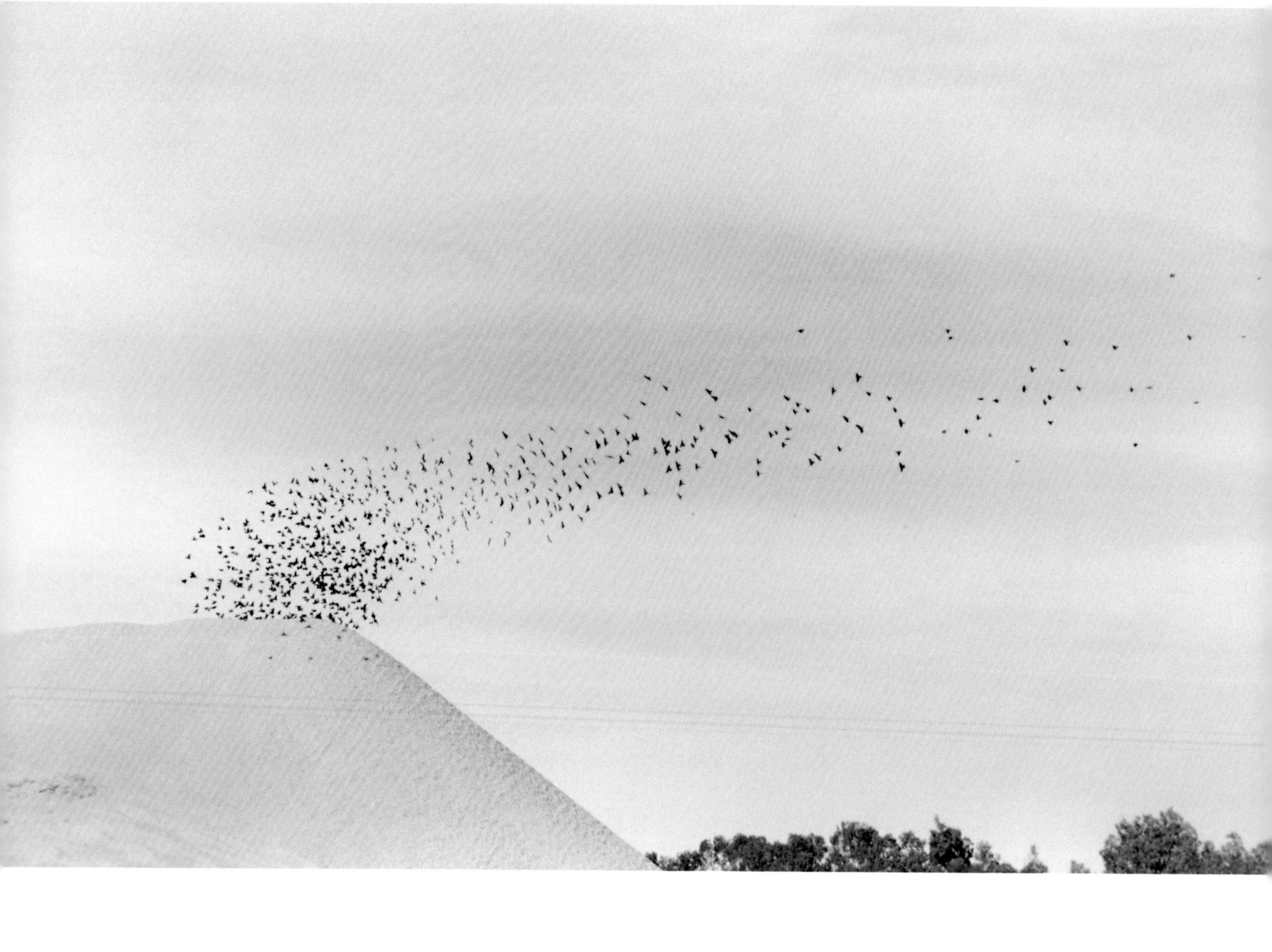

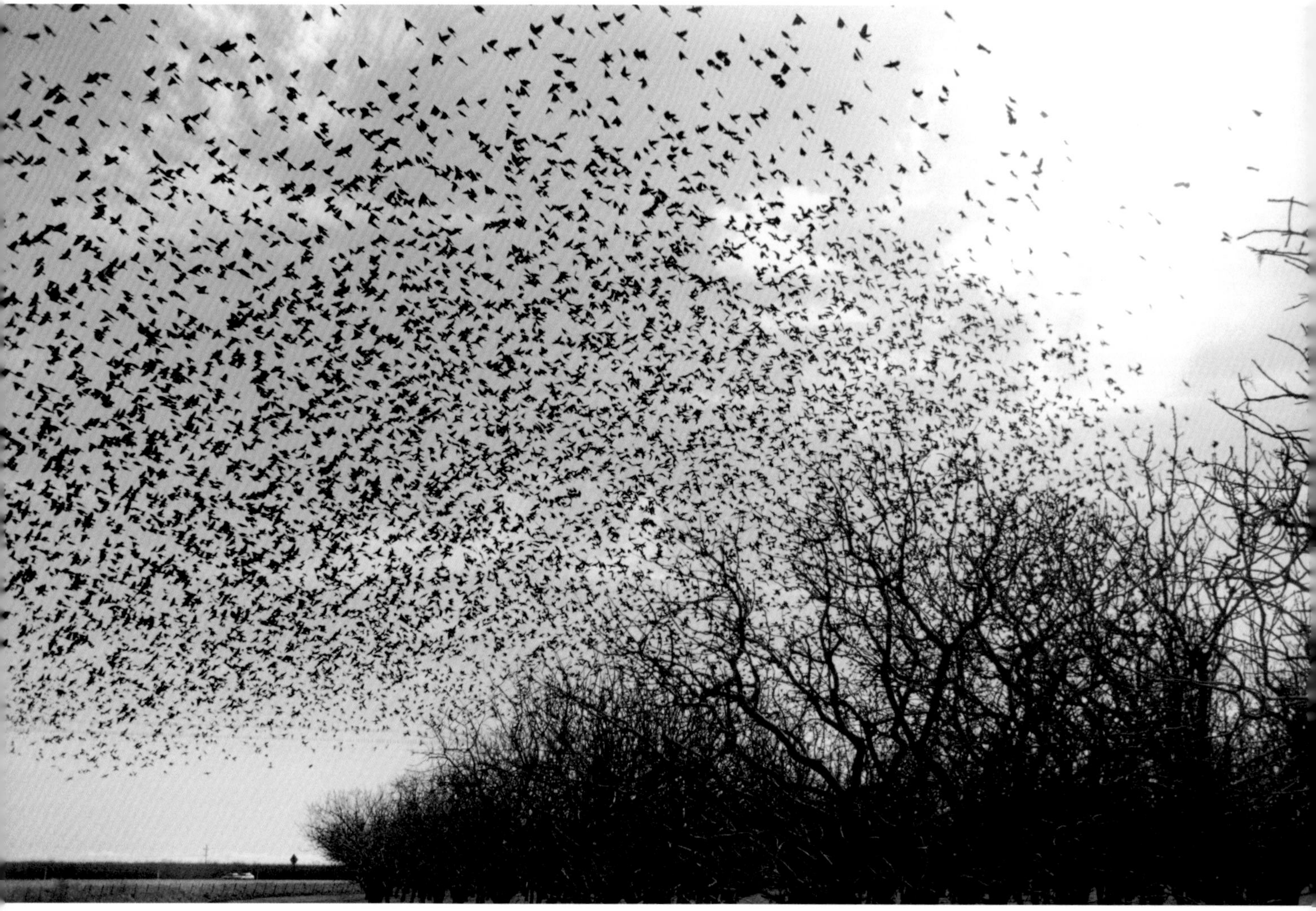

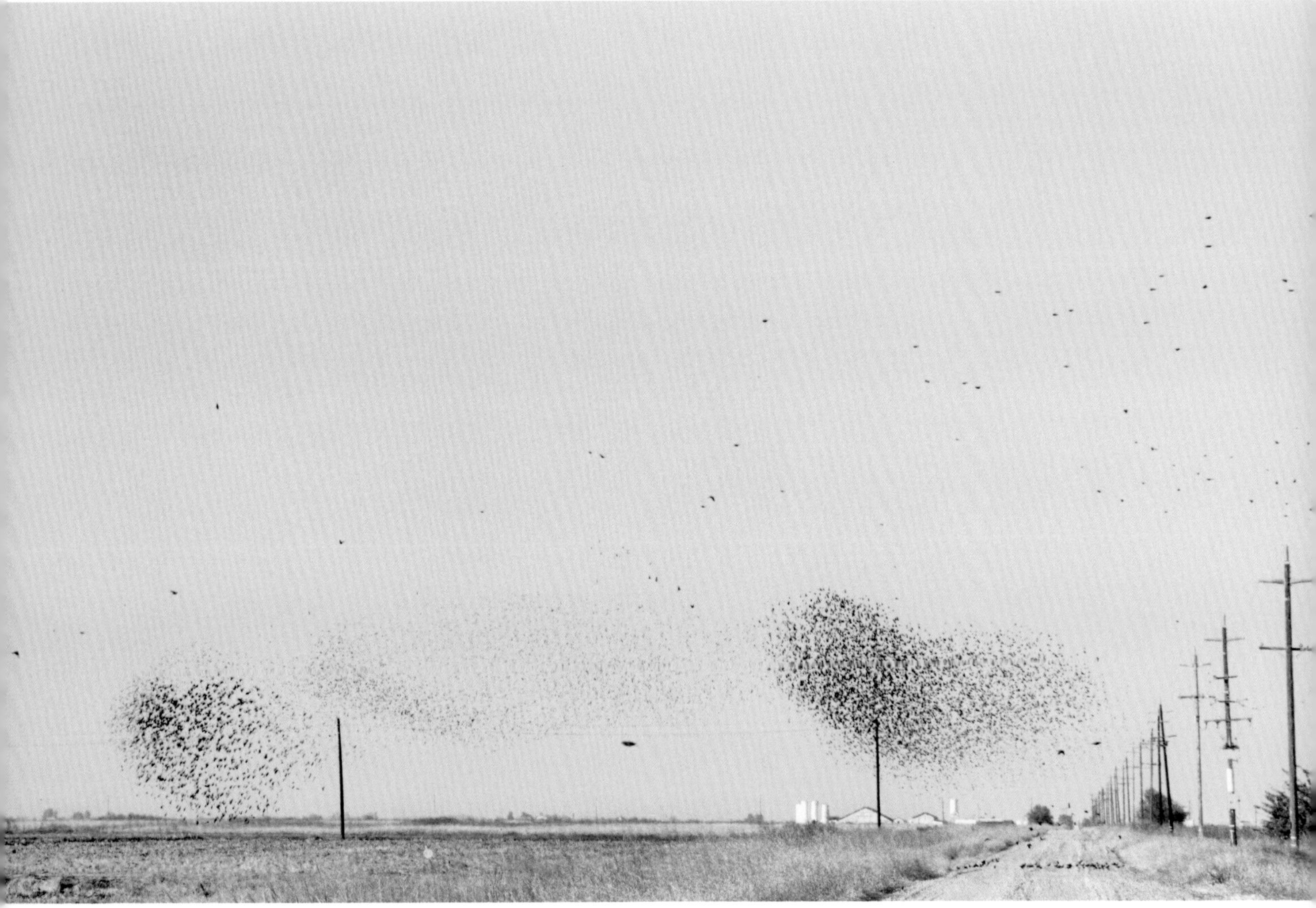

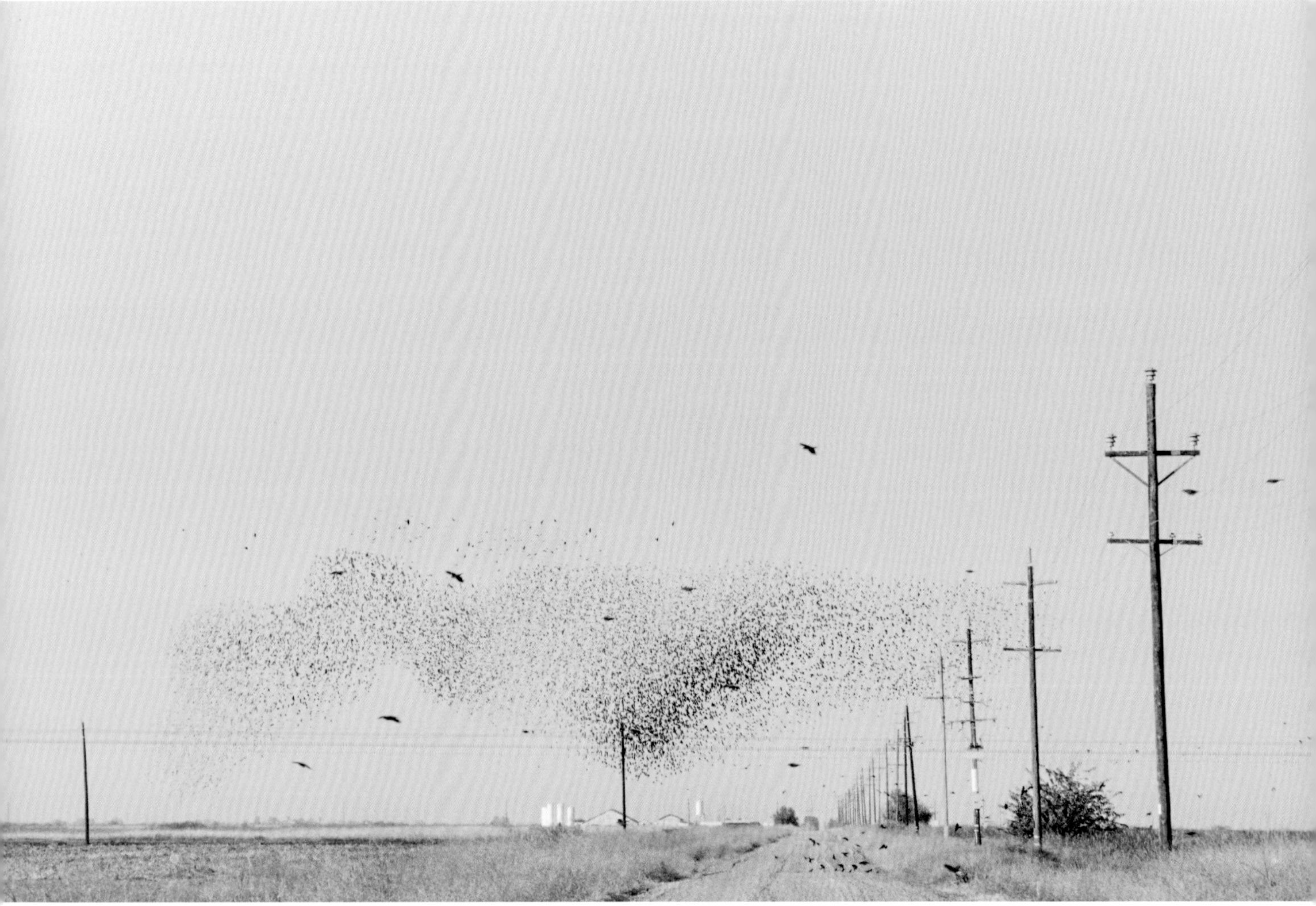

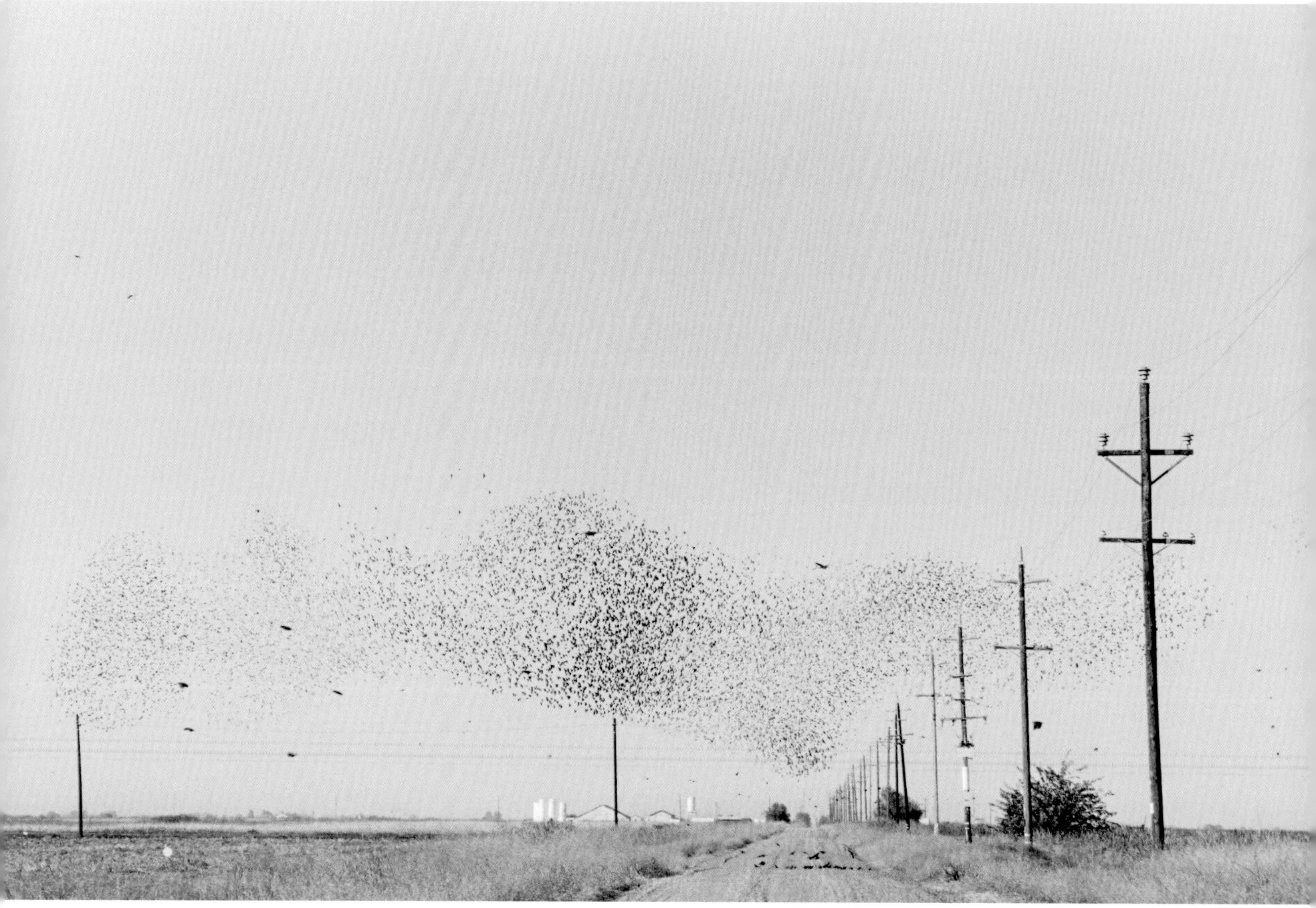

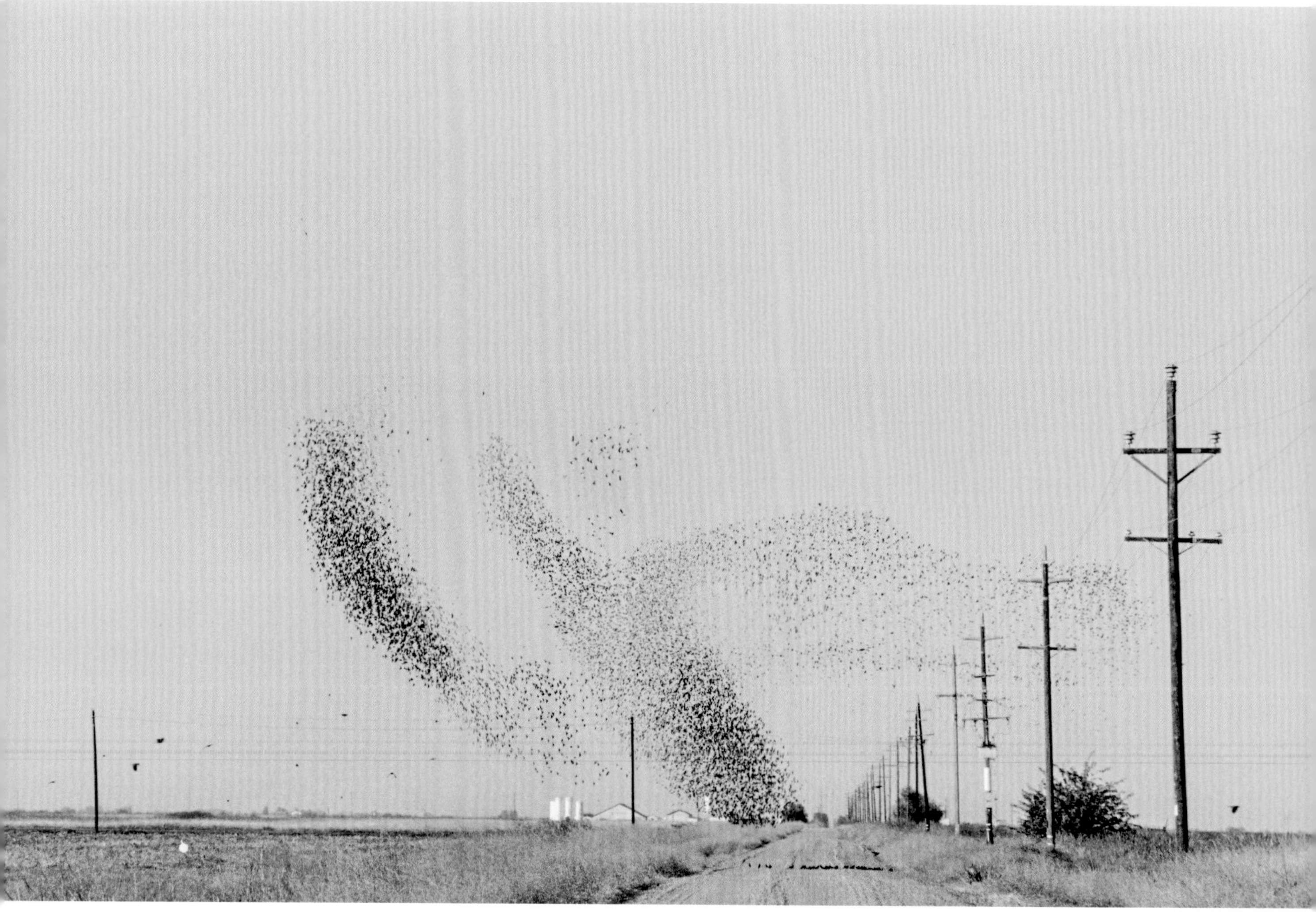

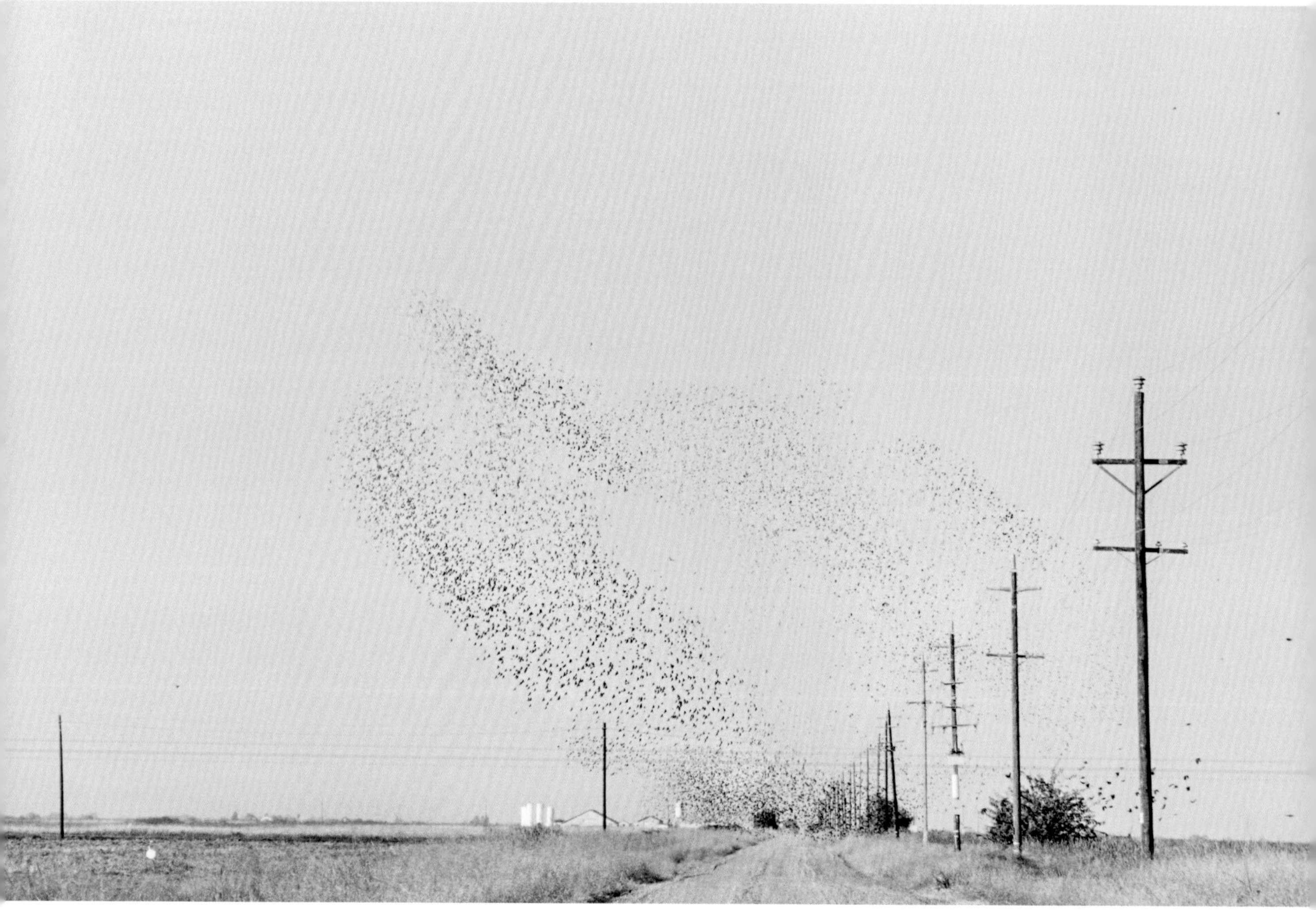

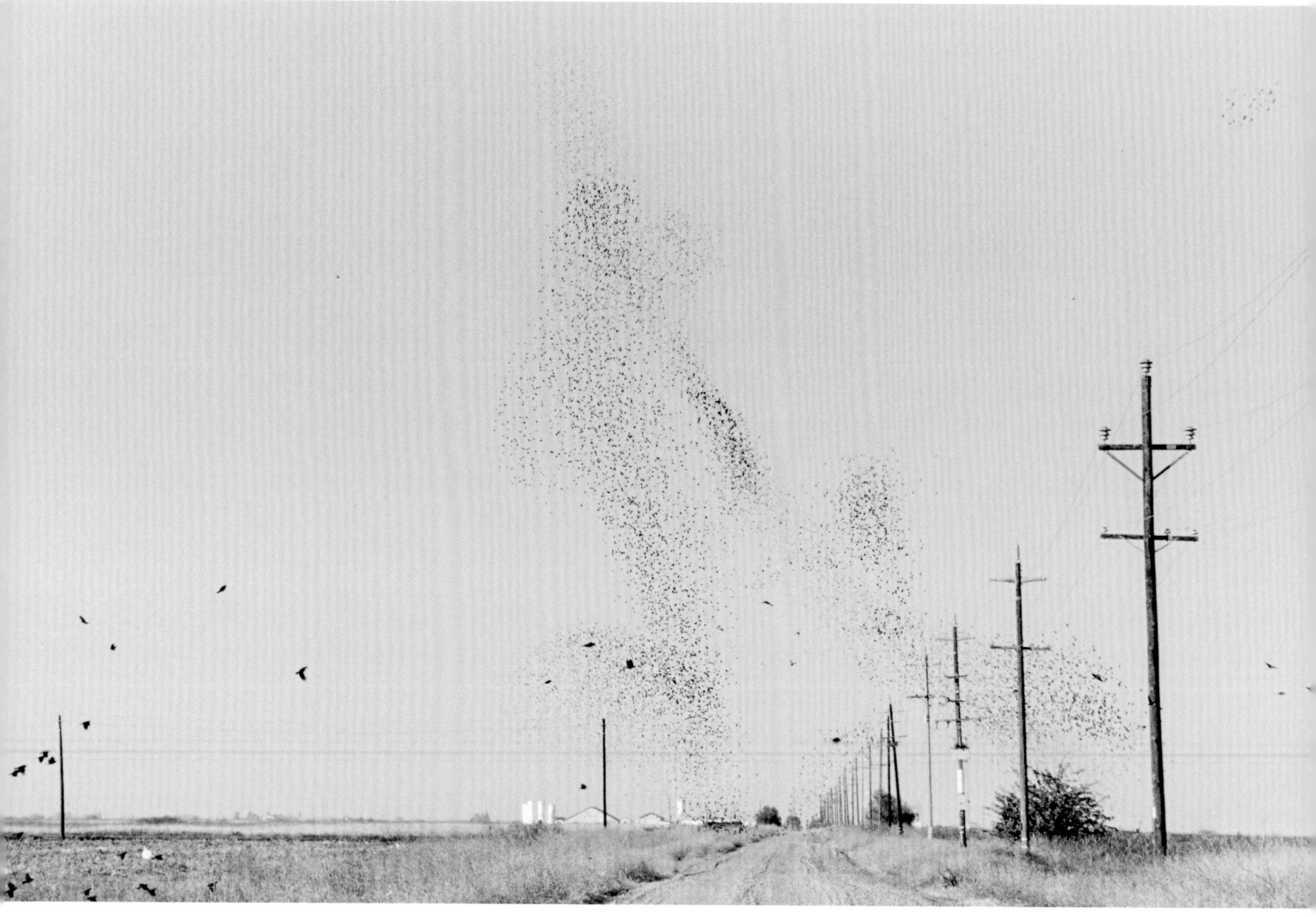

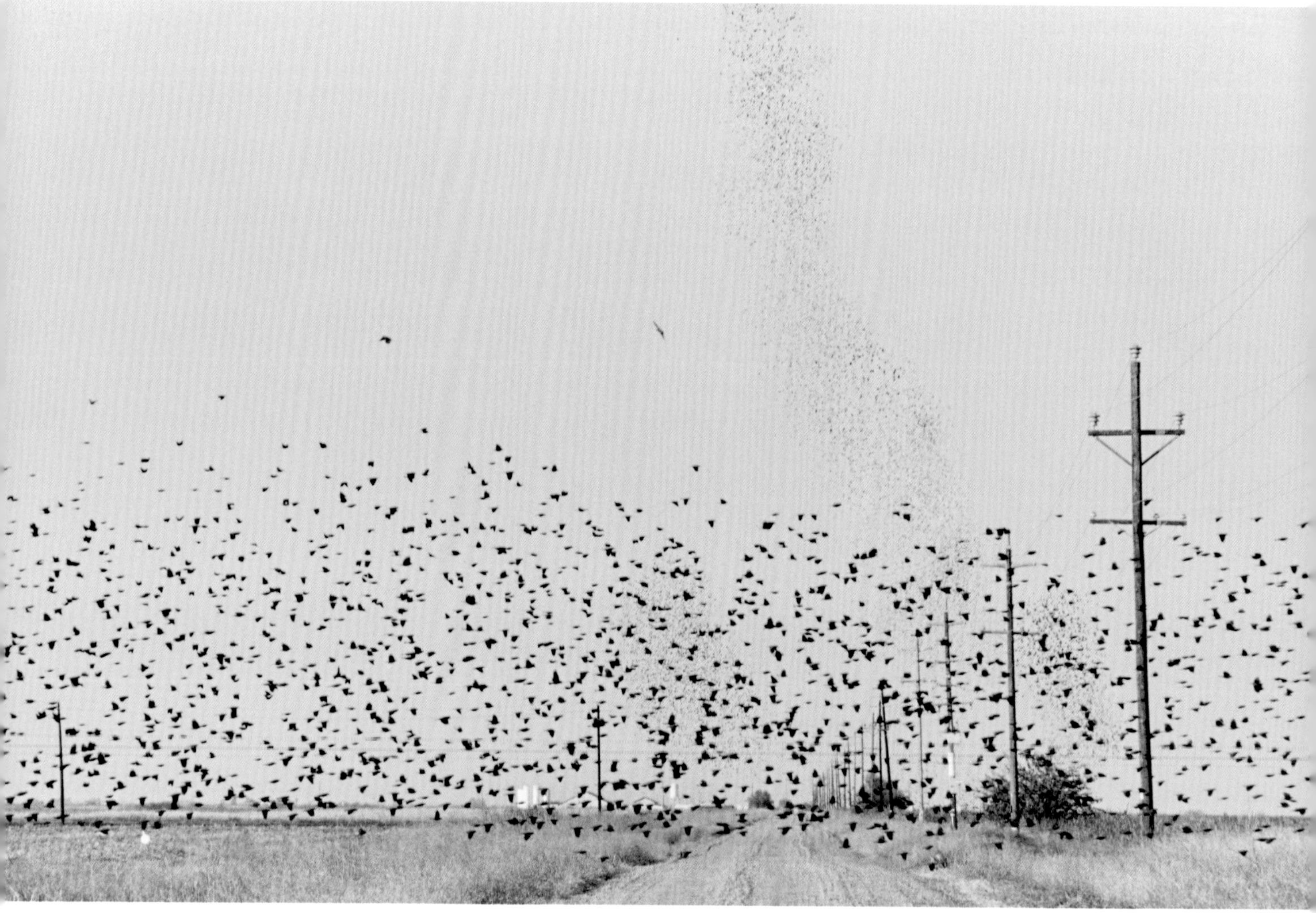

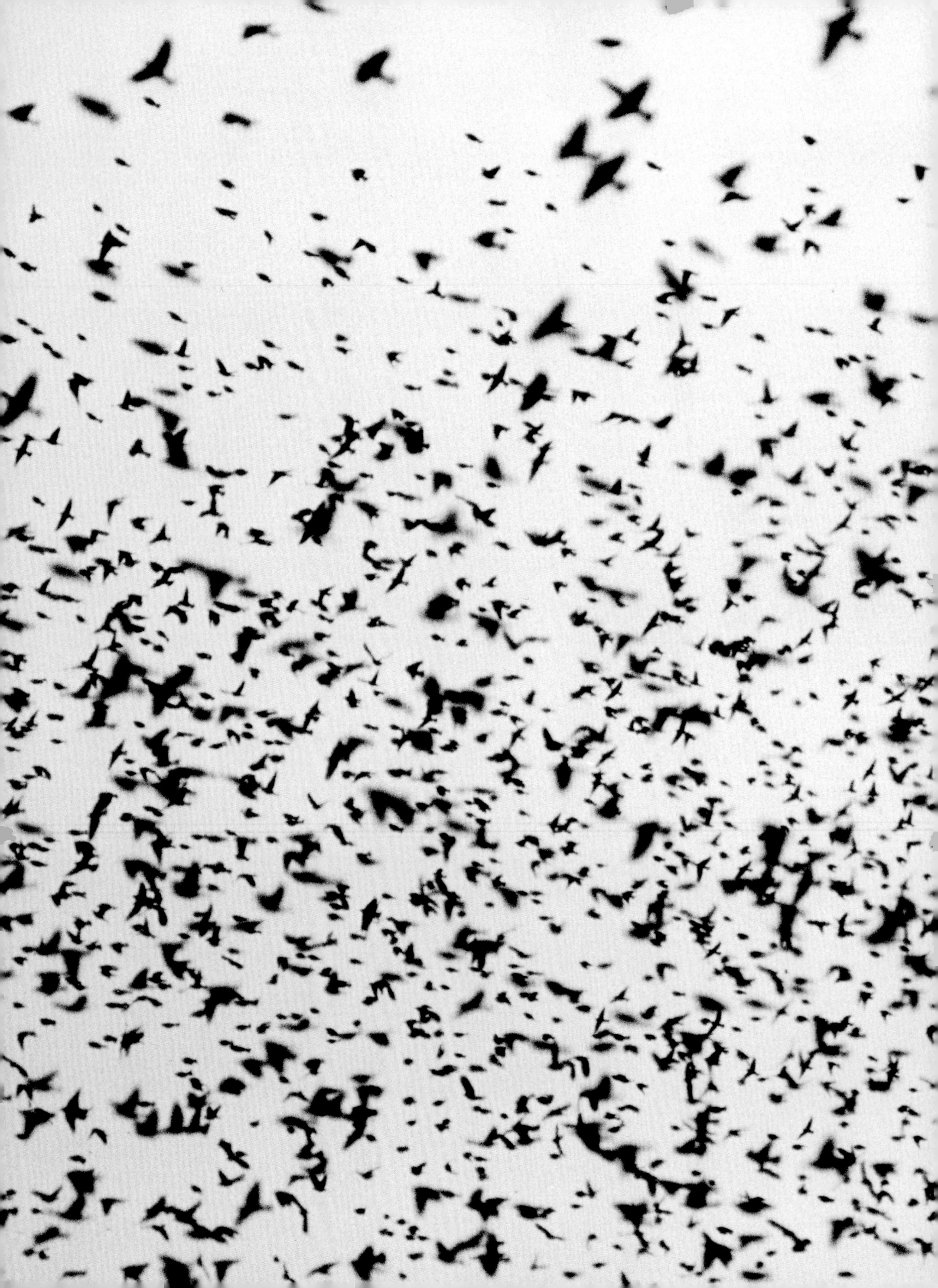

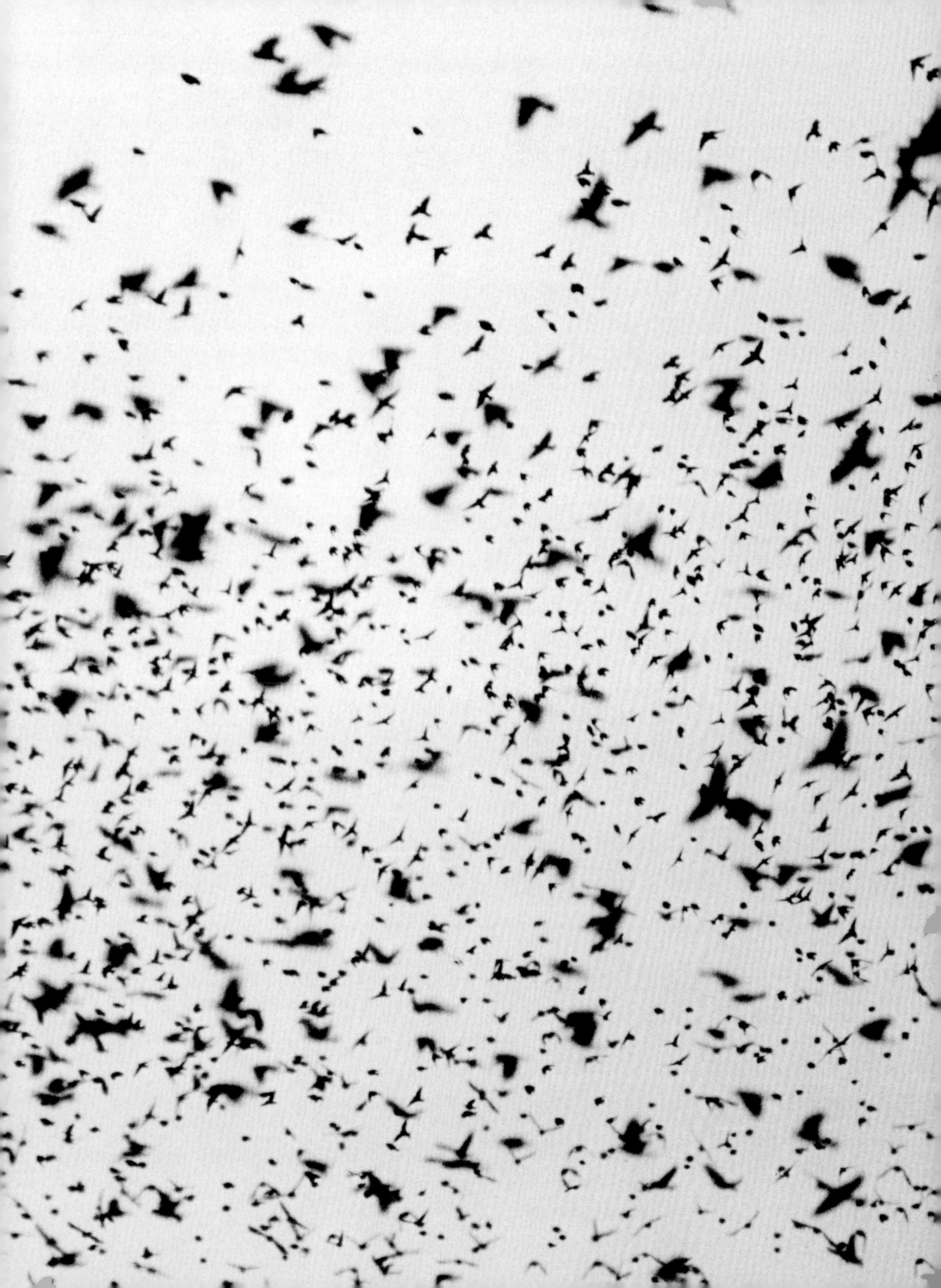

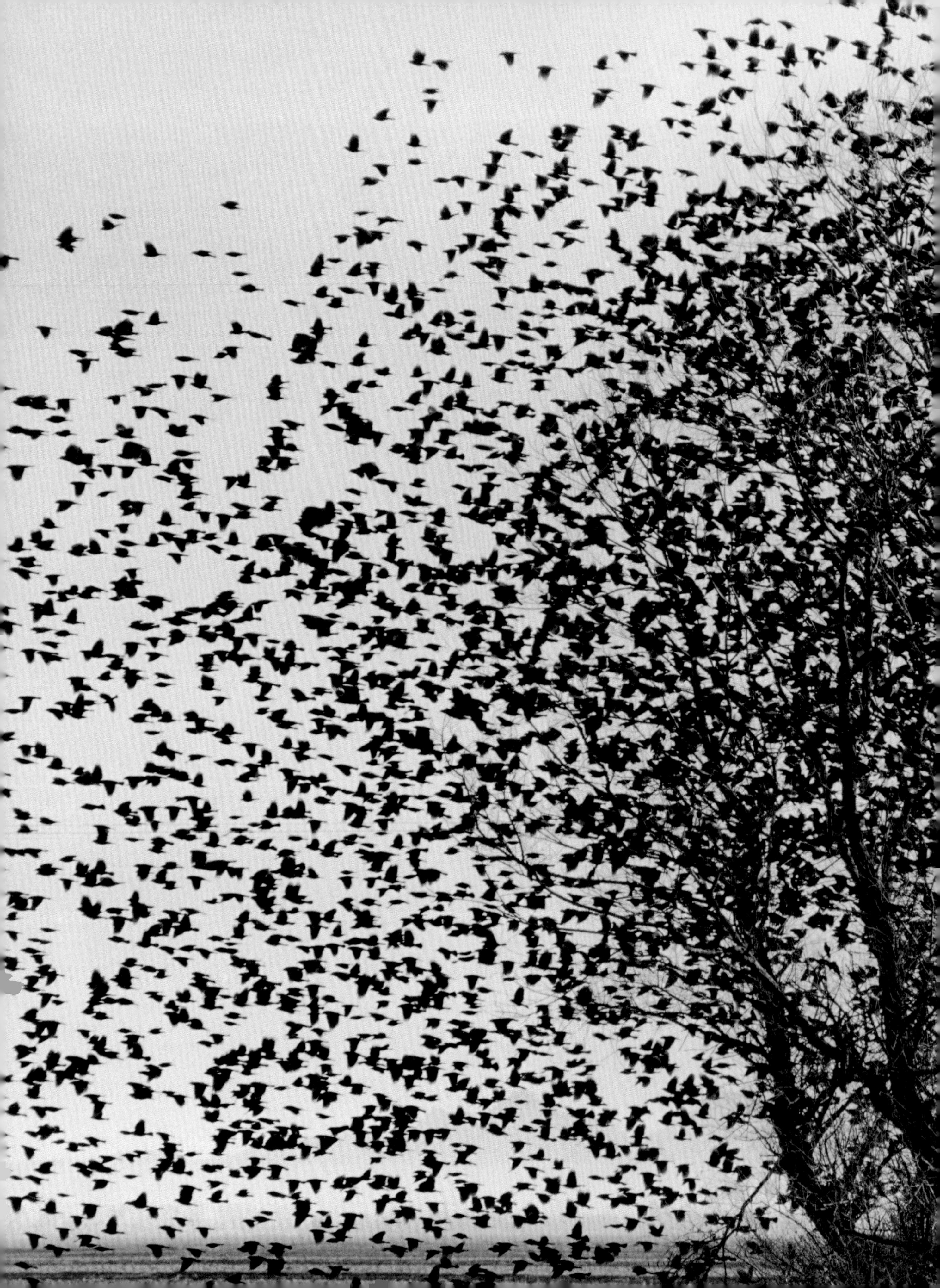

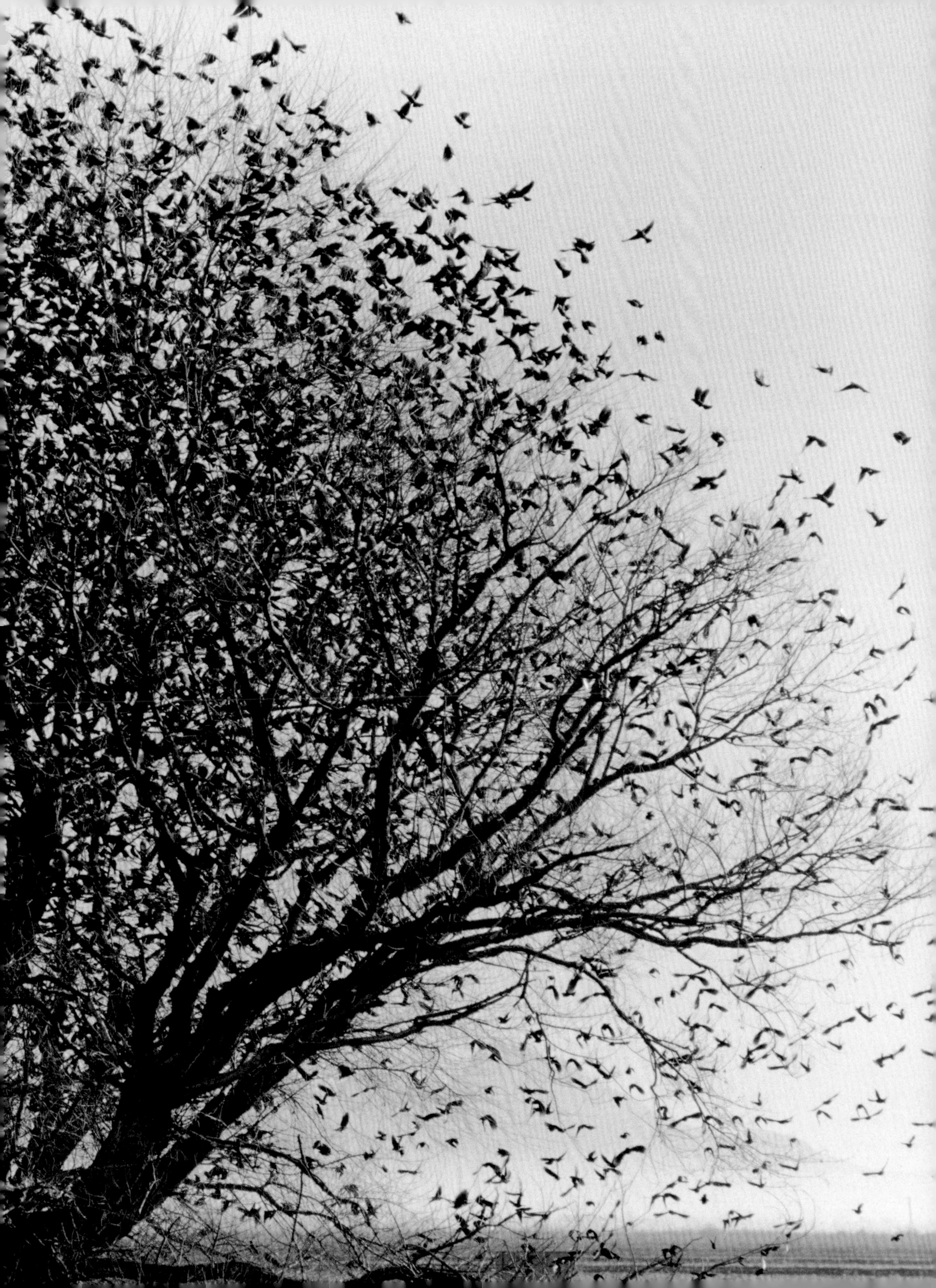

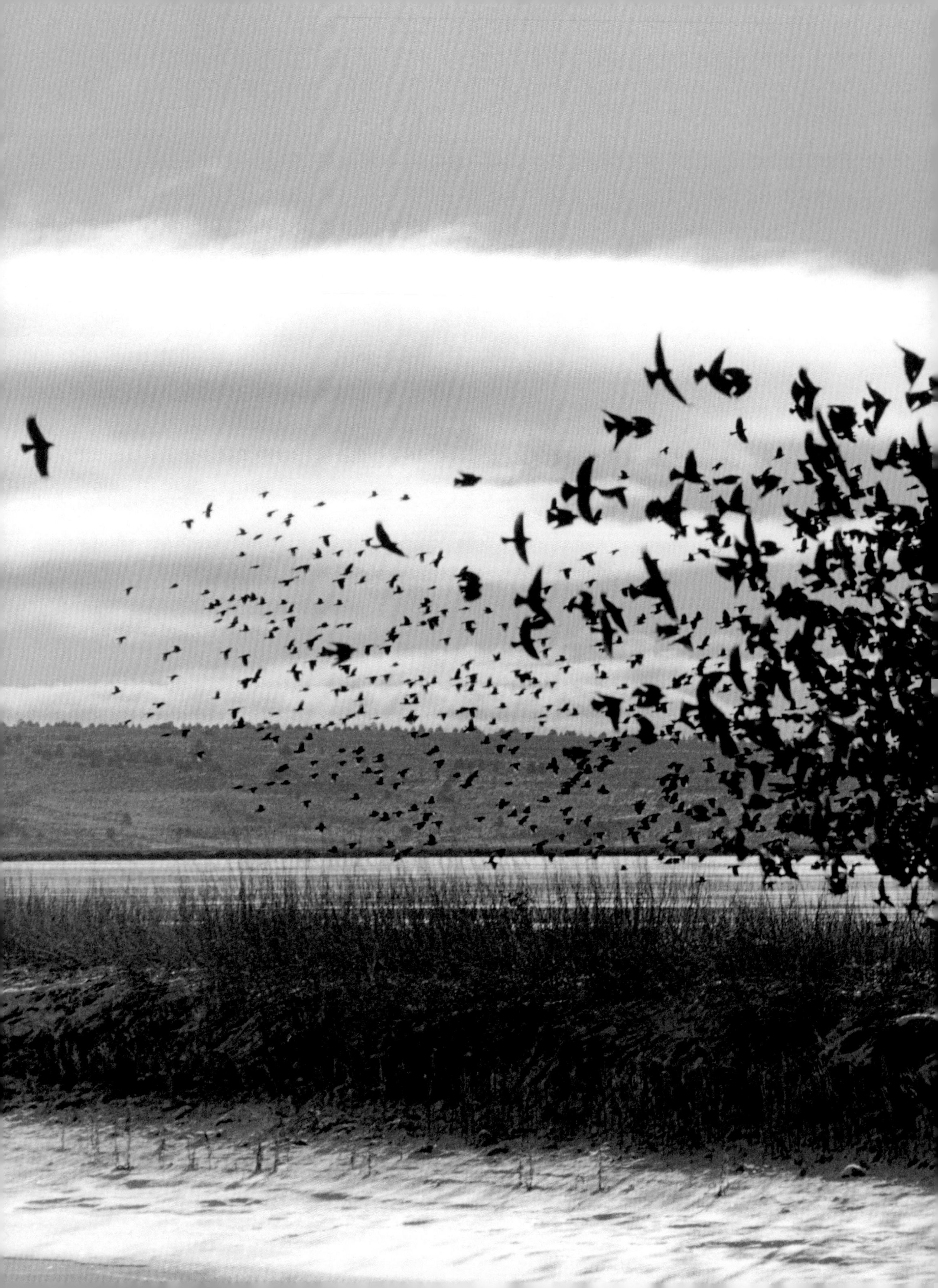

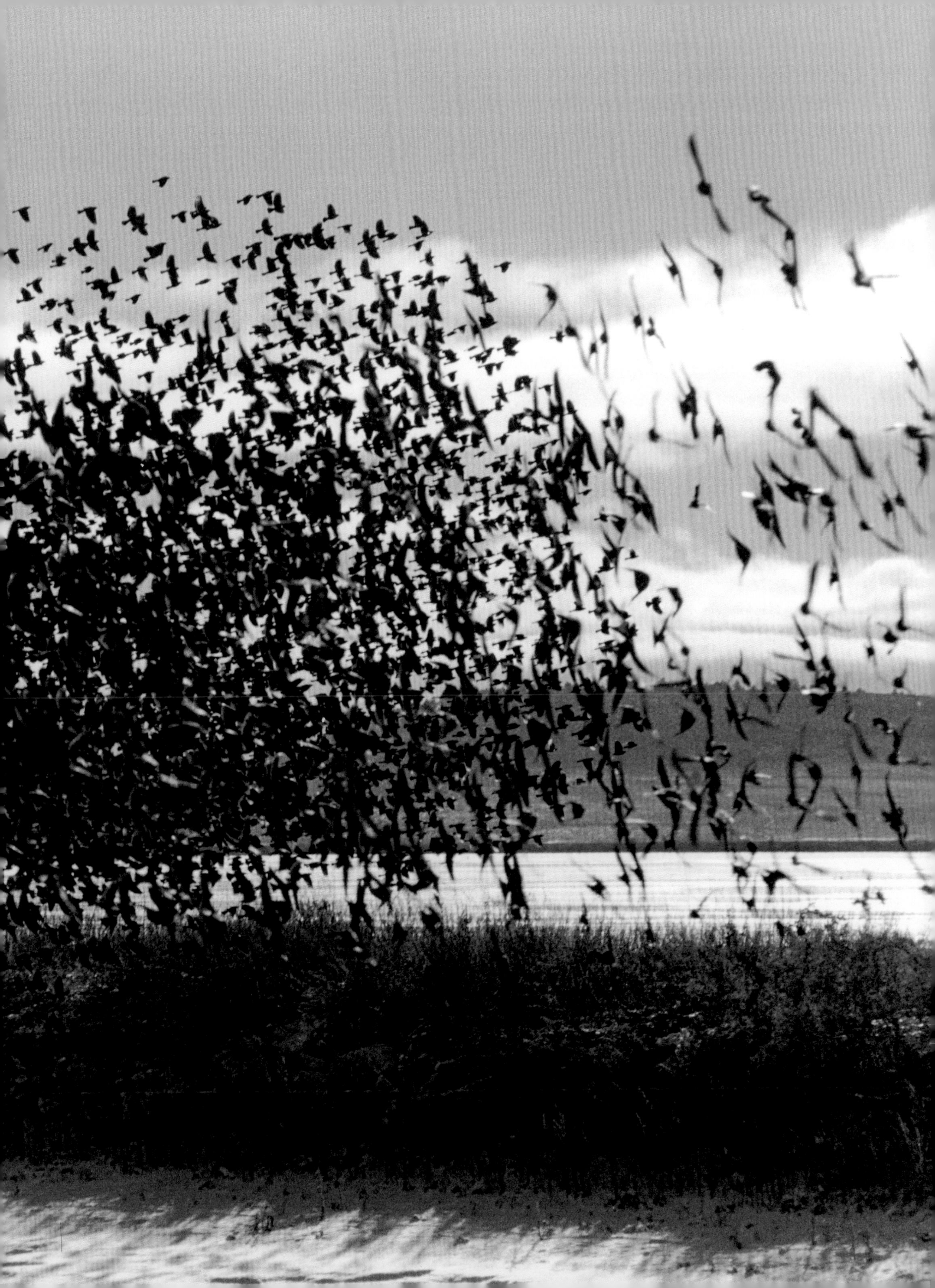

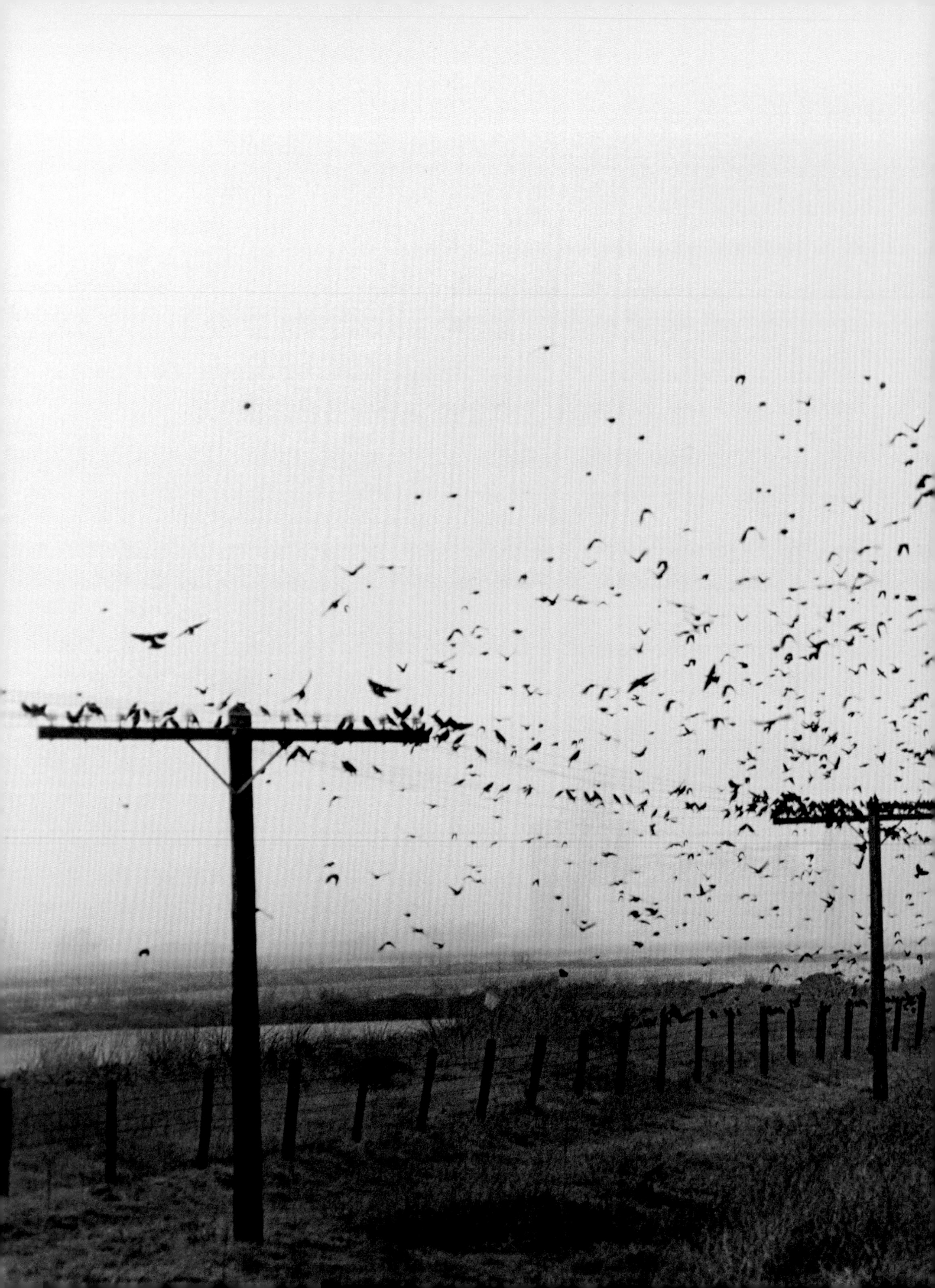

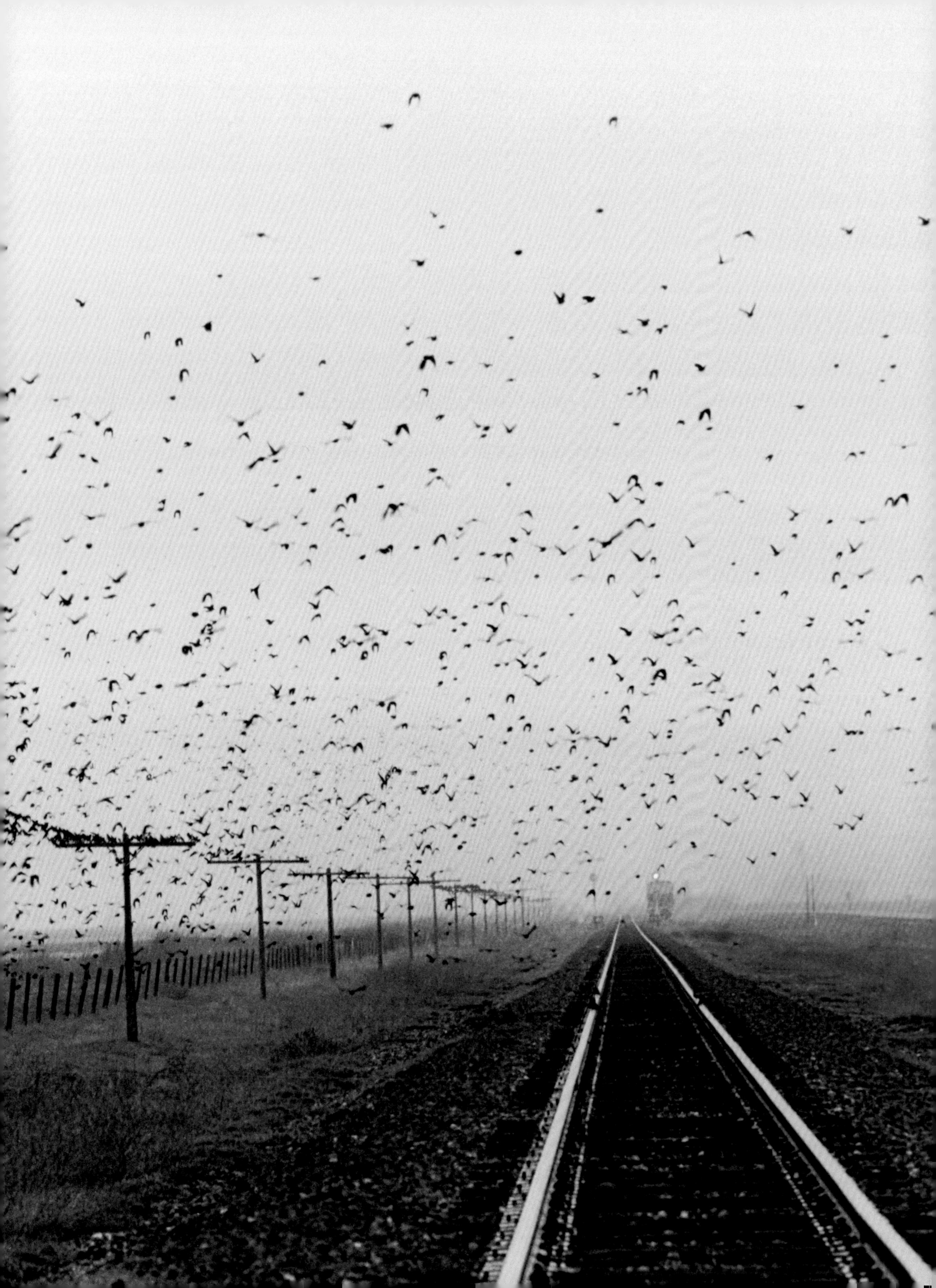

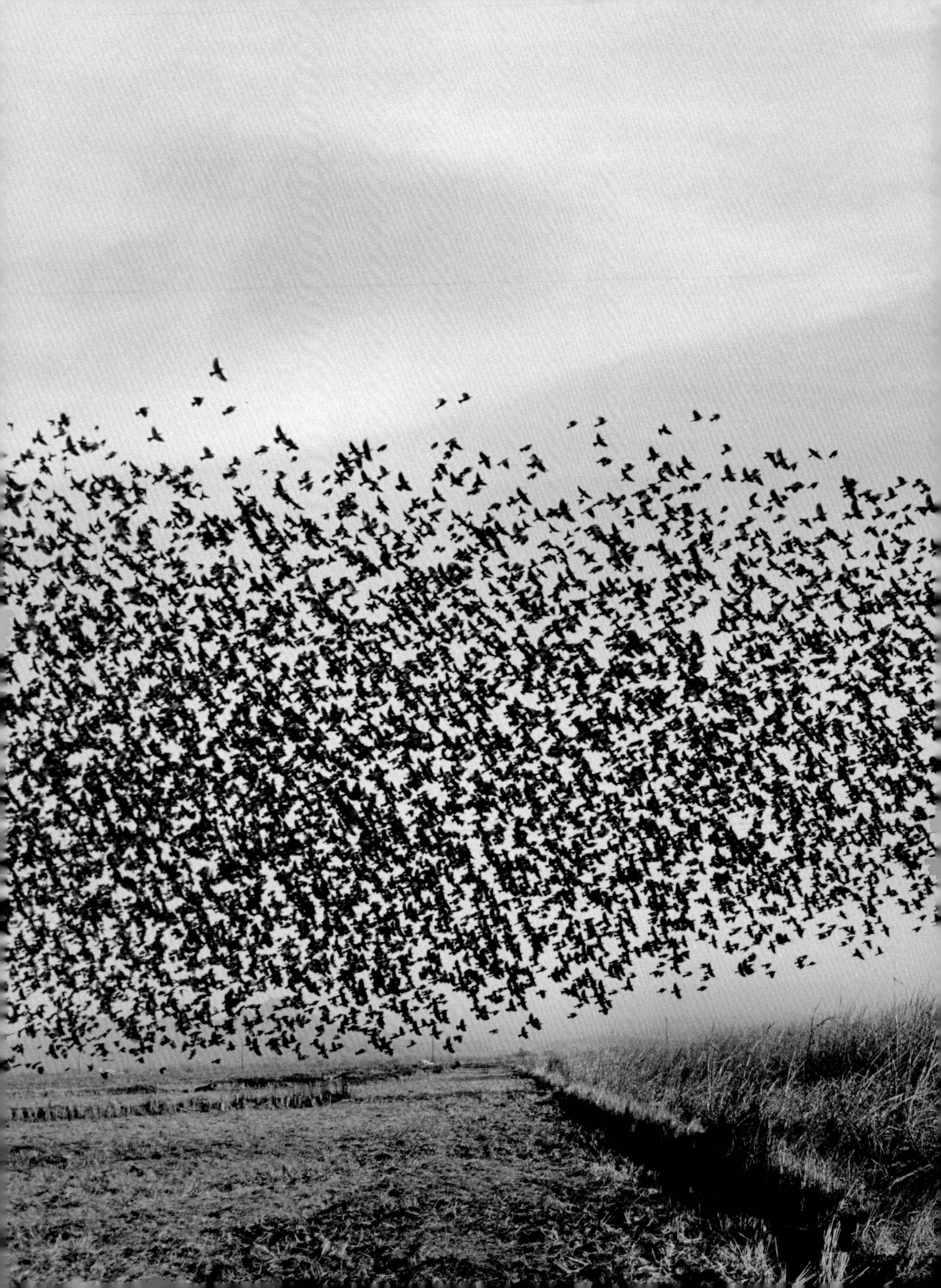

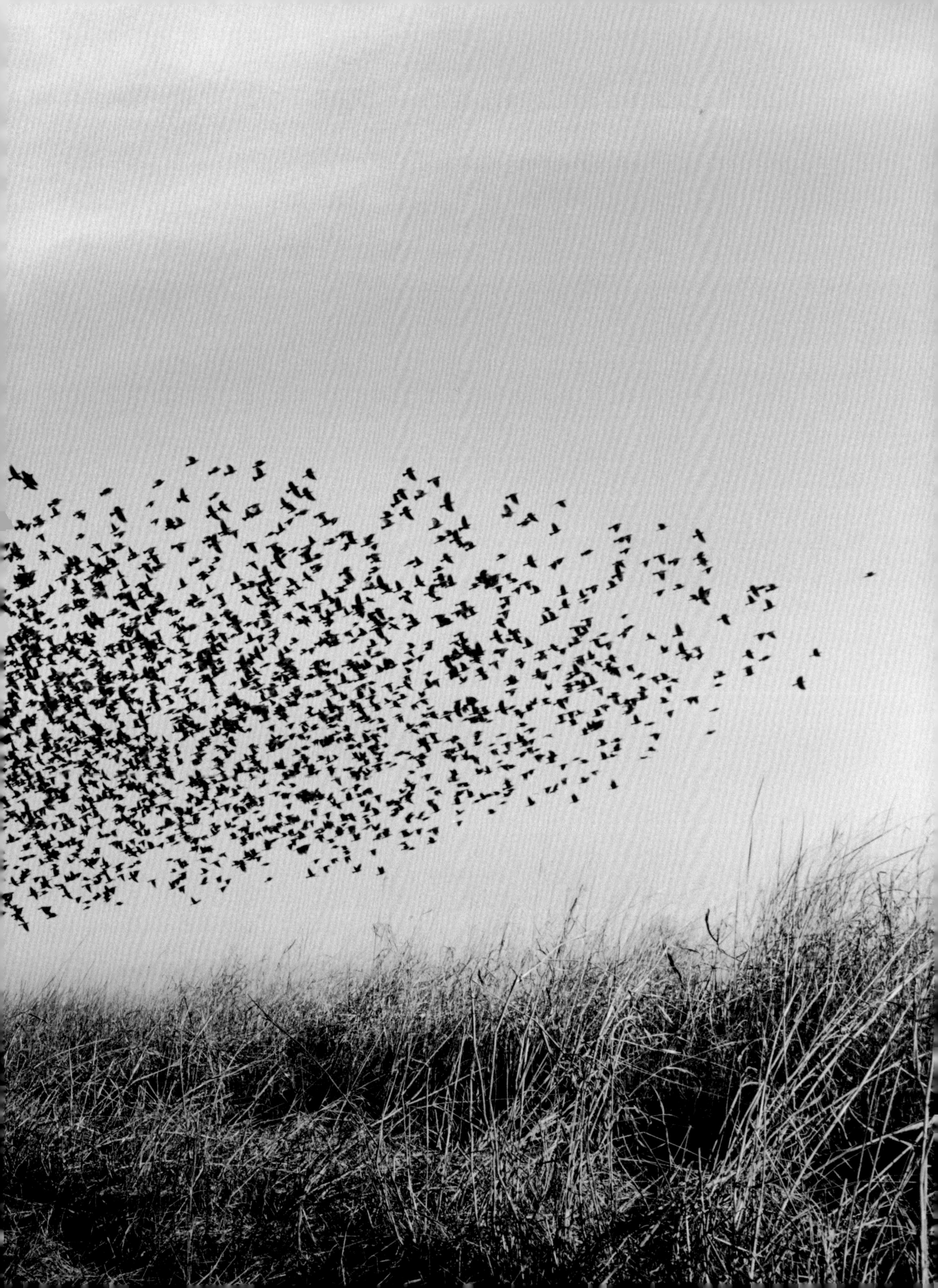

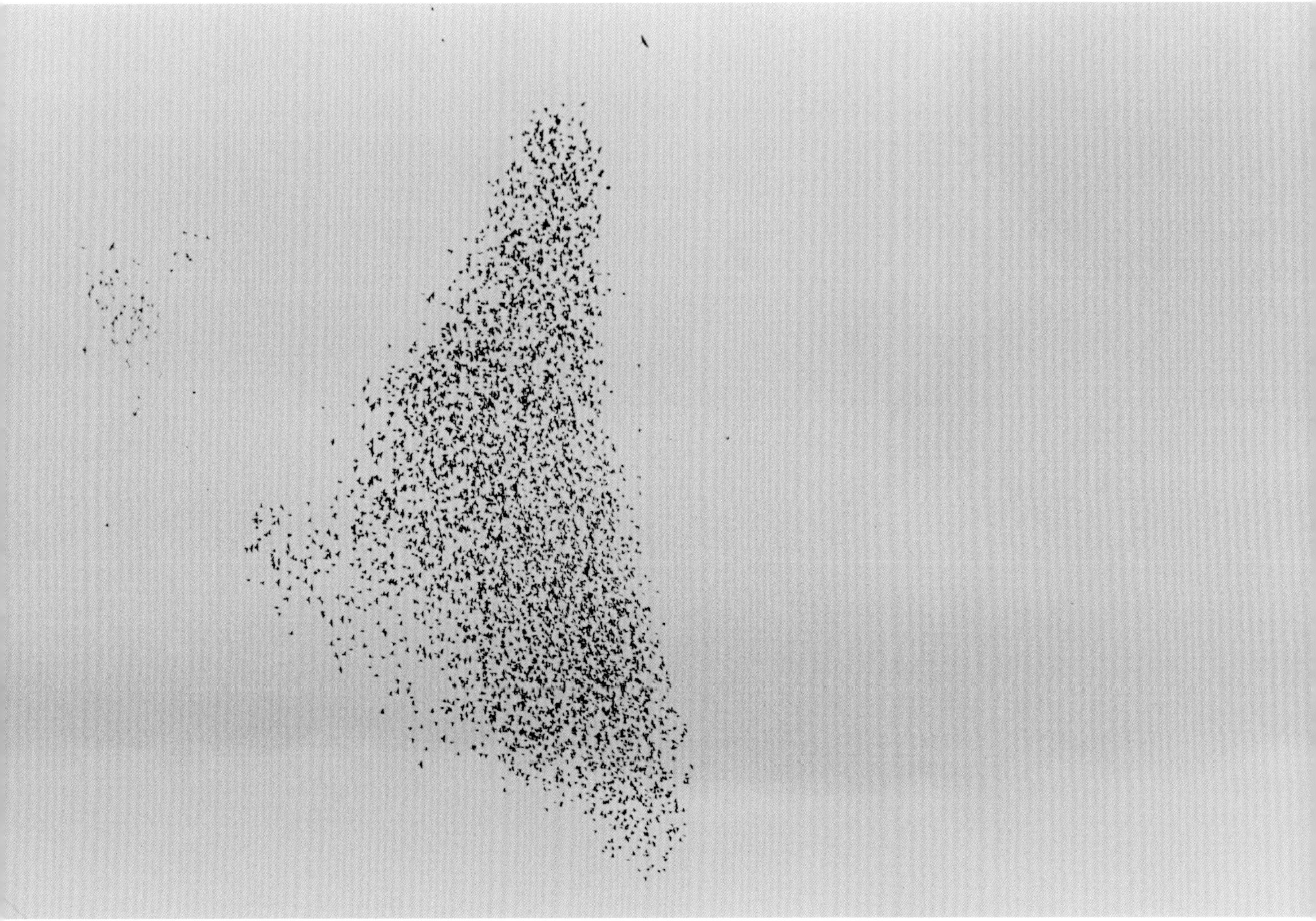

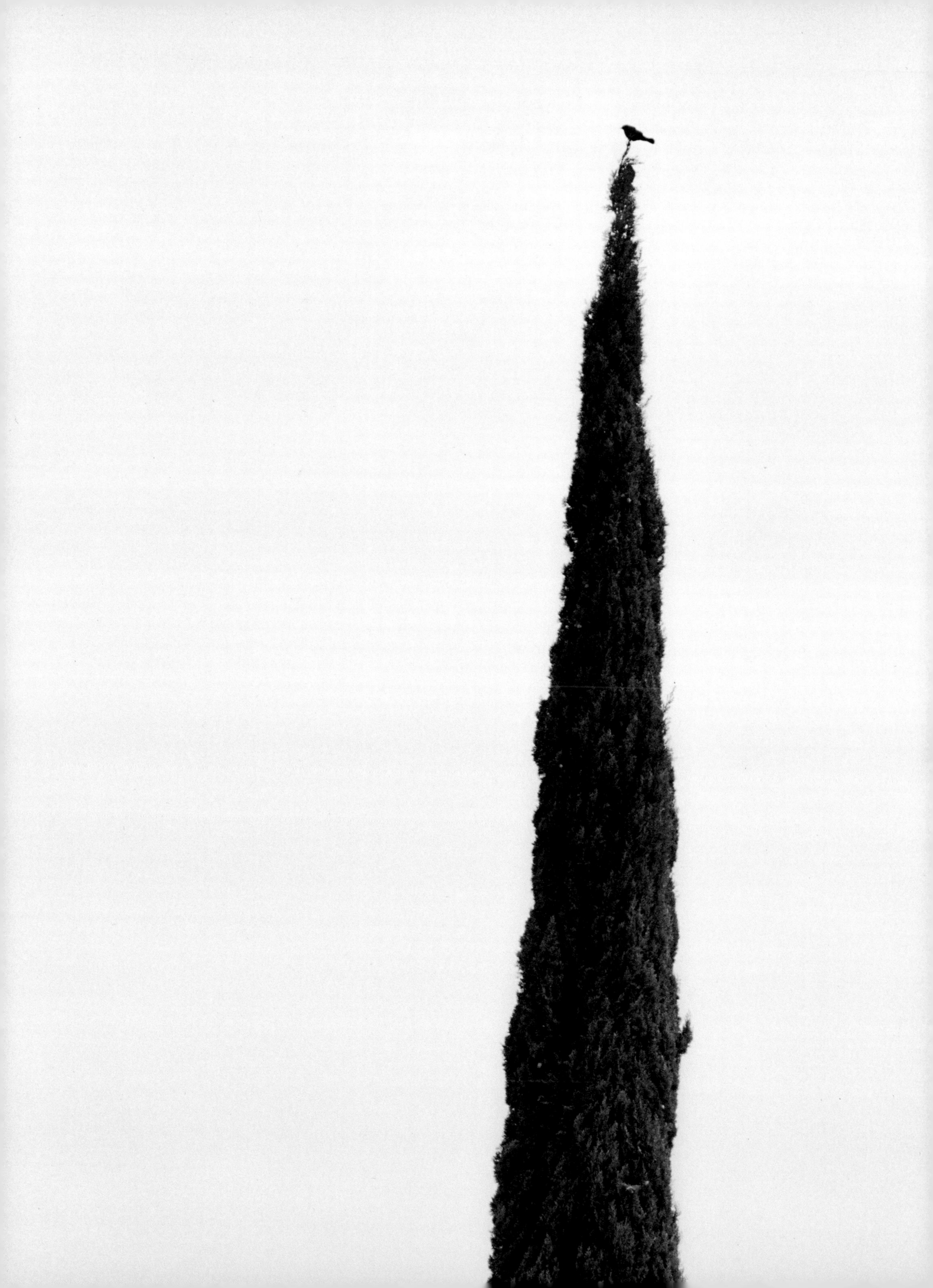

The Significance of Grouping in Blackbirds

Gordon H. Orians

Birds group for many reasons. To reproduce, males and females must associate at least briefly. Adults may remain together to nurture their offspring. They may join others to mob and chase away predators that come near their nests. They gather to drink at watering sites or to assemble at safe sleeping places. They may also associate with foraging individuals to find out where food is located. In some species, individuals cooperate to overpower prey that single birds would not normally be able to capture. Such benefits of grouping are important; otherwise birds would not assemble as often and under such varied circumstances as they do.

Nonetheless, there are inevitable disadvantages to living in groups. Individuals foraging together are likely to interfere with one another's ability to find food. A bird may unknowingly search for food where one of its associates has just removed prey. Predators easily find large groups. Lone individuals can hide; groups cannot! Disease-causing organisms can more readily find and infect new victims among clumped members of a species than among solitary individuals. Long before our ancestors knew that micro-organisms existed, much less that they caused diseases, they realized that isolating individuals helped reduce the spread of infections.

Because grouping has inherent disadvantages, there is always tension among group members over details of their associations. How close together should they be? Should additional group members be welcomed? Is it better to associate with some individuals more than with others? Should individuals of other species be accepted or expelled? Should individuals that appear sick be attacked? Answers to these and other questions depend on environmental circumstances, the existing size and composition of groups, and the status and needs of individuals. Continual behavioral interactions among group members occur as each individual acts to increase the benefits it derives from being in the group while decreasing its costs.

These complex interactions exist because grouping behavior will not evolve unless individuals that associate closely with others have a greater chance of survival and leave more offspring than those individuals that avoid such associations. Why individuals that associate with others to produce and care for offspring are more successful than those that remain celibate is obvious. Why individuals benefit from joining large flocks is not as self-evident. Solitary individuals can find food and seek safe resting places at night, thereby reducing the likelihood that they will contract diseases. Many birds, in fact, lead solitary lives, coming together only for the purpose of reproduction. What factors determine these striking differences?

Most North American species of blackbirds feed and roost at night in large flocks during autumn and winter when they are not breeding. The sizes of some of these flocks are astounding. Many winter roosts in the southern United States harbor more than a million individuals. Observers have estimated that one Texas roost contained fifty million birds; another had twenty-five million. Red-winged blackbirds are the most abundant species in most of the large roosts, but they are joined by large numbers of common grackles, brown-headed cowbirds, Brewer's blackbirds, starlings (not blackbirds), and smaller numbers of other blackbirds—rusty blackbirds, tricolored blackbirds, yellow-headed blackbirds, boat-tailed and great-tailed grackles, and red-eyed cowbirds.

These enormous roosting assemblages tell us that safe places to spend the night are rare. To find food, many of those birds undertake round trips of more than a hundred miles. Even though they do so only once each day, they expend large amounts of energy and time they could otherwise spend foraging.

In winter, blackbirds of all species eat mostly seeds. Plants produce those seeds during summer, after which seed production stops. As a result, during autumn and winter, foraging blackbirds steadily deplete their food supply. They first deplete the seeds close to their roosts, gradually extending the area they exploit. Late winter, when seeds are scarce and insect populations have not yet increased, is usually the time of the greatest food shortage.

Even though it is advantageous to roost together in particularly safe places, birds in those huge aggregations need not forage in flocks. Birds can easily leave the roosts and fly off to forage individually. Foraging flocks are much smaller than the groups of birds that assemble to roost in one place, suggesting that being in an enormous flock is not helpful when the objective is to find food. Nevertheless, most North American blackbirds gather in flocks, some of them large, while foraging during autumn and winter. What benefits do they gain that more than compensate for the costs they inevitably incur? We cannot ask birds directly why they flock, but their behavior does give us clues.

For example, if foraging in flocks does indeed provide protection from predators, then we should be able to observe several interesting things. First, that predators have less success when attacking flocks than when attacking solitary individuals. The more individuals in a flock, the greater the likelihood that there is always at least one individual on the look-out to detect an approaching predator before it gets dangerously close. Secondly, if there are many birds in a flock, each individual can spend most of its time foraging without there being times when no other bird is alert. Both of these patterns have been observed in bird flocks. Birds in flocks detect predators faster than solitary individuals do. The bigger the flock, the less time individuals spend looking up, enabling them to spend more of their time foraging.

Thirdly, if flocking provides protection, we would also expect birds to form larger flocks when foraging in more dangerous locations or when they are aware that predators are nearby. Experimenters have flown trained hawks over foraging birds. When they detect the hawk, small birds quickly assemble into larger flocks, especially during times of food shortage. When they feel safe, foraging birds spread out, thereby reducing interference from other birds while they are foraging.

Thus, abundant evidence shows that foraging in flocks provides protection from predators, that those benefits increase with flock size up to a point, and that birds can adjust their flocking behavior according to current estimates of risk. Red-winged blackbirds foraging in open fields are vulnerable to predators. There is no place to hide; a black bird is conspicuous against a brown background. In contrast, meadowlarks (also blackbirds) with their striped brown backs are difficult to spot in grasslands if they crouch on the ground. Meadowlarks do not form large flocks and often forage solitarily during winter.

Protection from predators appears to be the major factor that favors foraging in flocks during autumn and winter, but birds also have ways of reducing the inevitable disadvantages that come with being close to others. To reduce interference while foraging they move closer together or spread farther apart depending on their assessment of risk. They also have ways of reducing foraging interference and reducing the chance that they will seek food where another bird has just extracted prey. You may have watched a flock of blackbirds "roll" across a field. This rolling movement results from an interesting pattern of movement by flock members. When an individual finds itself at the back of the flock, it flies over its flock-mates and settles at the front of the flock. It forages there in a small area until it again finds itself at the rear of the flock. It then flies back to the front. This behavior maintains flock cohesion while giving each individual a place to forage that was not previously visited by other flock members. And it need not fly far to do so.

Blackbirds foraging in fields are all seeking the same food. In forests, however, many different kinds of food are available and they are found in different places—under bark, on small and large branches, on twigs, leaves, and flowers, and flying through the air. In forests, foraging birds are also vulnerable to predators, especially avian predators that perch concealed among foliage and branches. Having many eyes alert for danger is highly valuable in forests and woodlands. Not surprisingly, many woodland birds form mixed-species feeding flocks during the non-breeding season. Members of such flocks typically include trunk and branch foragers (woodpeckers, nuthatches), gleaners of small branches (tanagers, vireos) and leaves (warblers, chickadees), and pursuers of flying insects (flycatchers). In addition to serving as predator-spotters, these birds may disturb insects that are then more readily captured by individuals of other species. More eyes but less interference is the key to mixed-species foraging flocks in woodlands.

Birds in large flocks risk not only interfering with one another's foraging, but also bumping into one another as the flocks wheel through the air. This risk increases with the approach of a predator because then birds clump closer together. It is dangerous for a falcon to dash into a flock of birds because a collision with a bird, even a small one, is likely to injure its wing. Falcons attempt to separate a single individual from the flock and attack it. (Fish form dense aggregations when approached by predators for the same reason!)

But how do birds avoid bumping into one another when flying in dense flocks, and how do they achieve the striking and remarkably coordinated twisting and turning movements that so fascinate us? Some individual is inevitably at the front of a flock, and we humans, thinking of our own social lives, are quick to assume that there must be a leader who initiates departures, turns, and landings. But this easy assumption appears to be wrong. Analyses of flock movements clearly indicate that any bird in a flock can initiate a turn that subsequently propagates through the flock in a wave that radiates from the initiating individual. The analyses also show that the wave travels much too fast to be explained by the assumption that each bird watches and responds only to its nearest neighbor. The time it takes a bird to respond to a light flash, as measured in a laboratory, is about thirty-eight milliseconds. Turning waves propagate through a flock more than twice that fast. To account for the rapid propagation rate it is necessary to assume that birds see a wave approaching and are ready to turn before it reaches them. Wayne Potts, who carried out the analyses, dubbed this the "chorus line hypothesis" because that is how chorus line girls accomplish their coordinated turns.

Not surprisingly, computer experts have developed models that try to mimic real flock movements. To make successful predictions, models need only a few simple rules. 1. Steer to avoid crowding local flock-mates. 2. Steer towards the average heading of local flock-mates. 3. Steer to move towards the average position of local flock-mates. This suggests that the remarkably complex coordinated movements we observe may be the result of rather simple behavioral rules and that there is no single leader. However, the sense of wonder we feel when watching the flocks that are so strikingly captured in Lukas Felzmann's photos is in no way diminished by the knowledge that complex patterns can be generated by simple mechanisms.

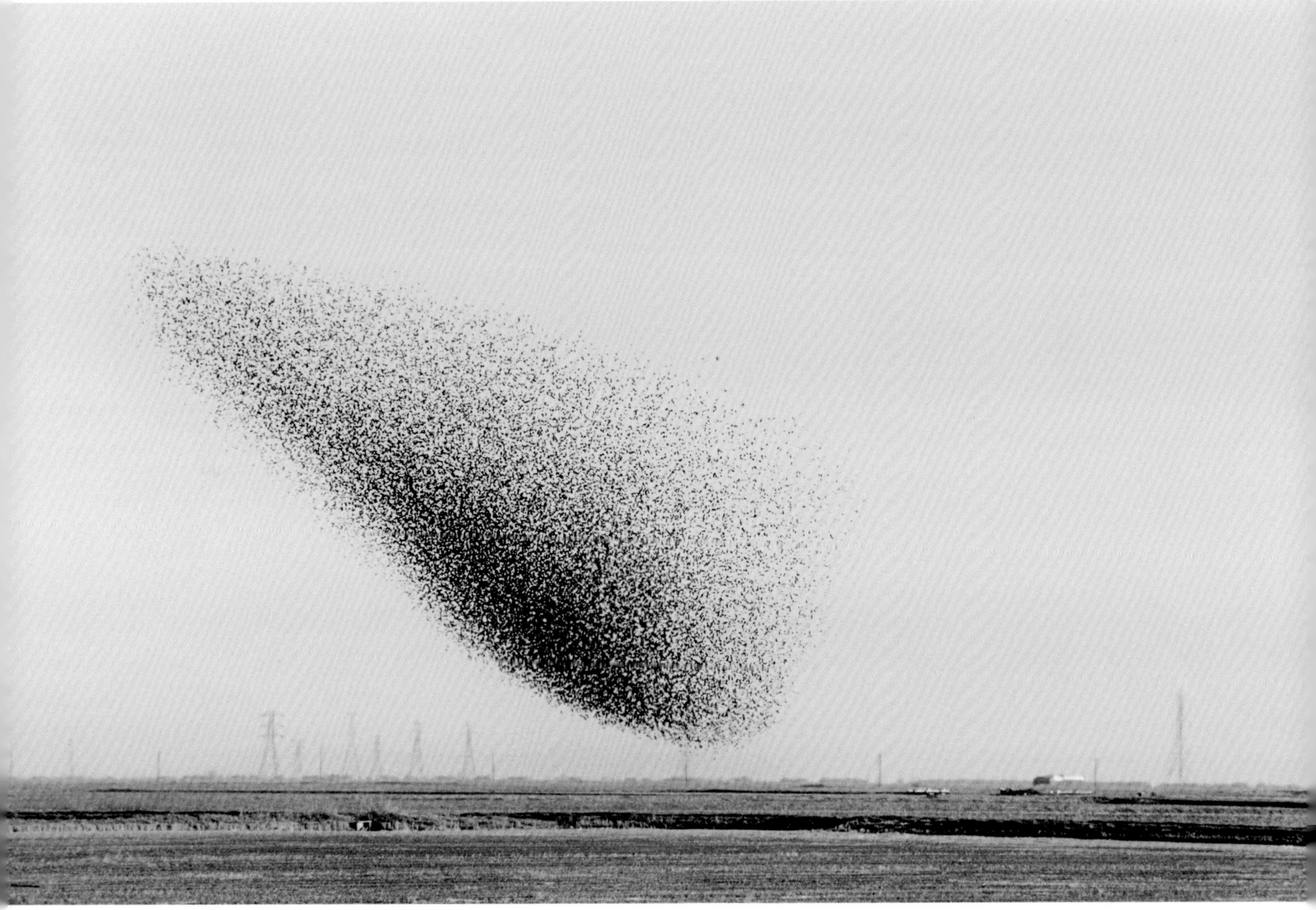

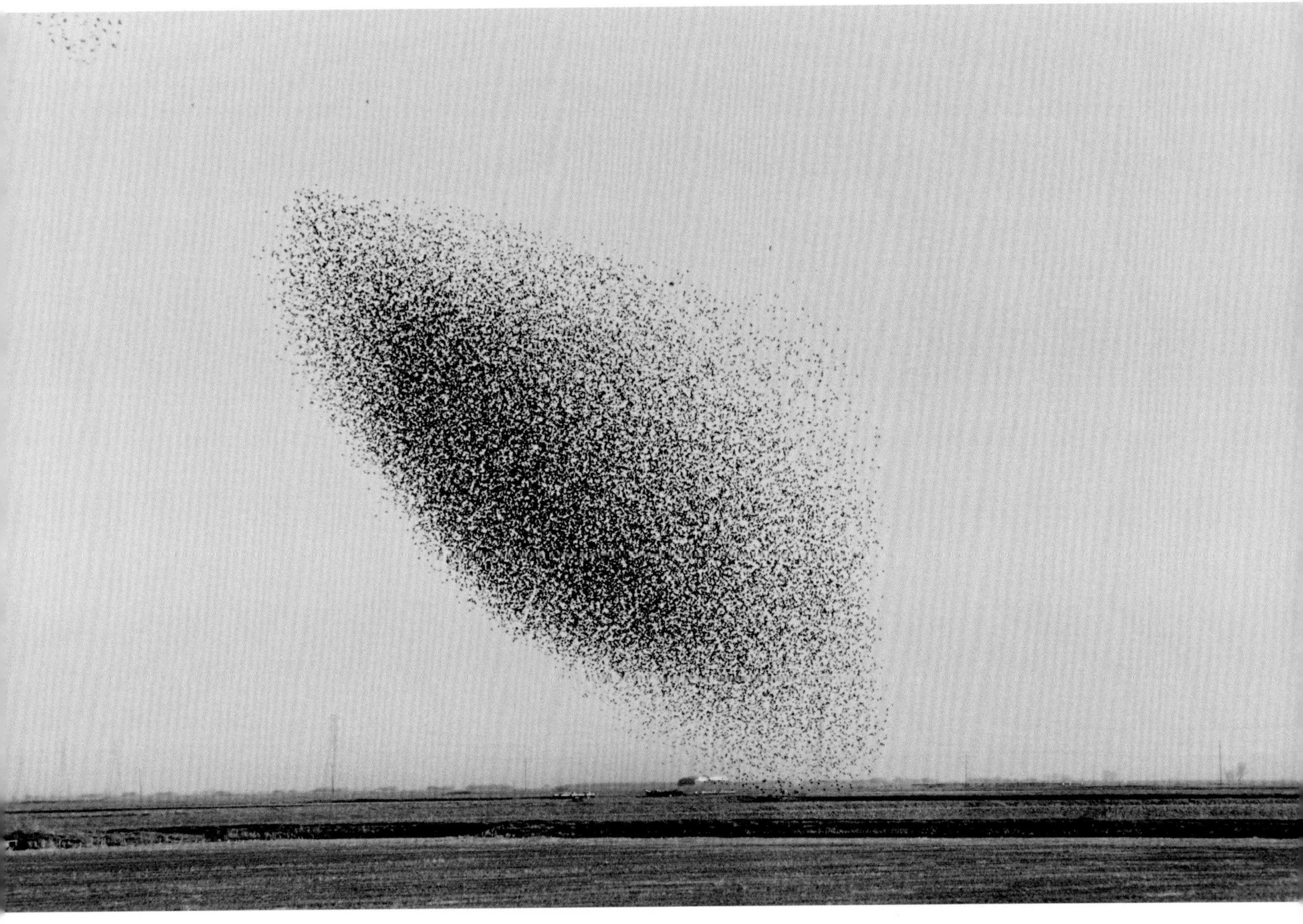

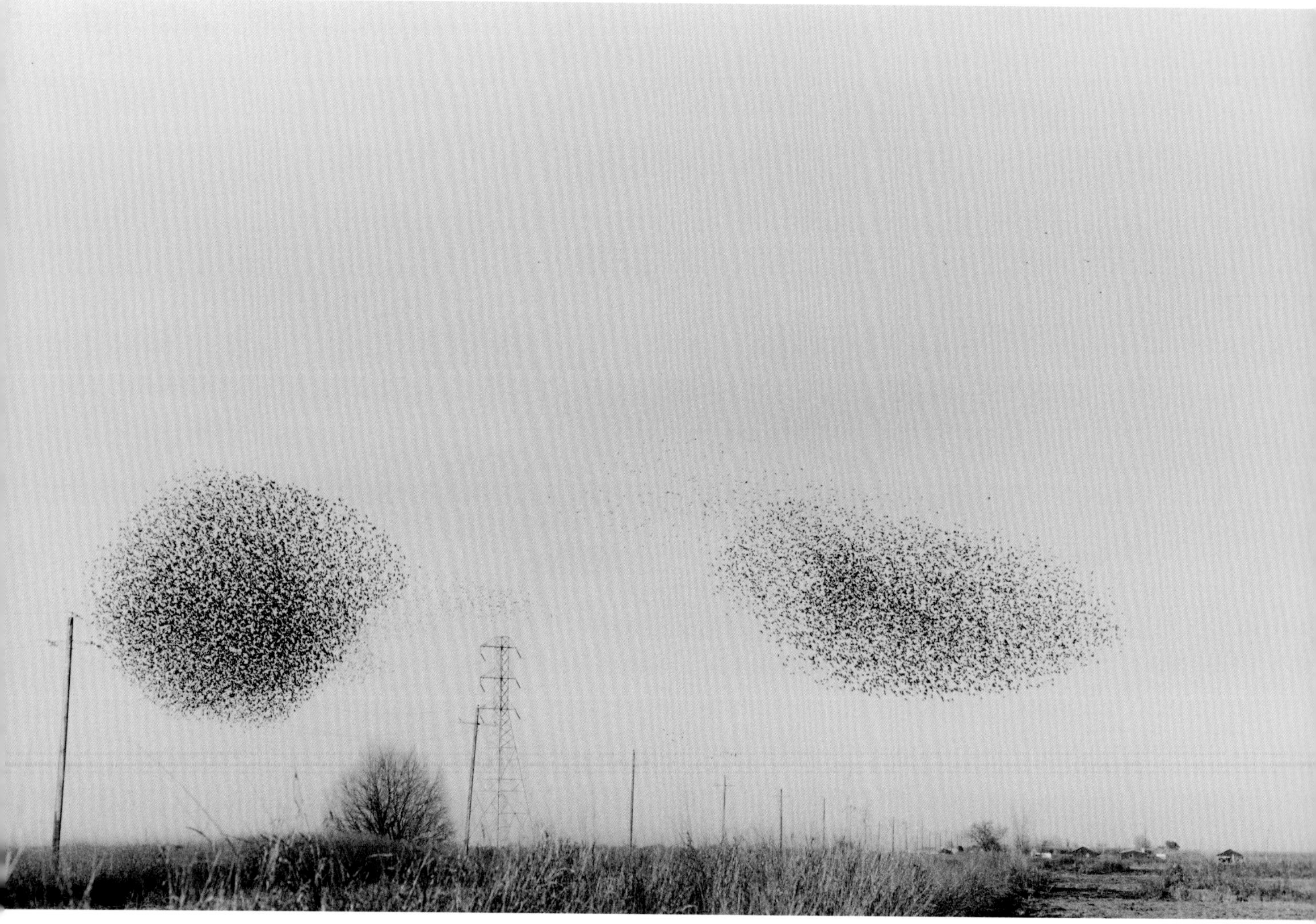

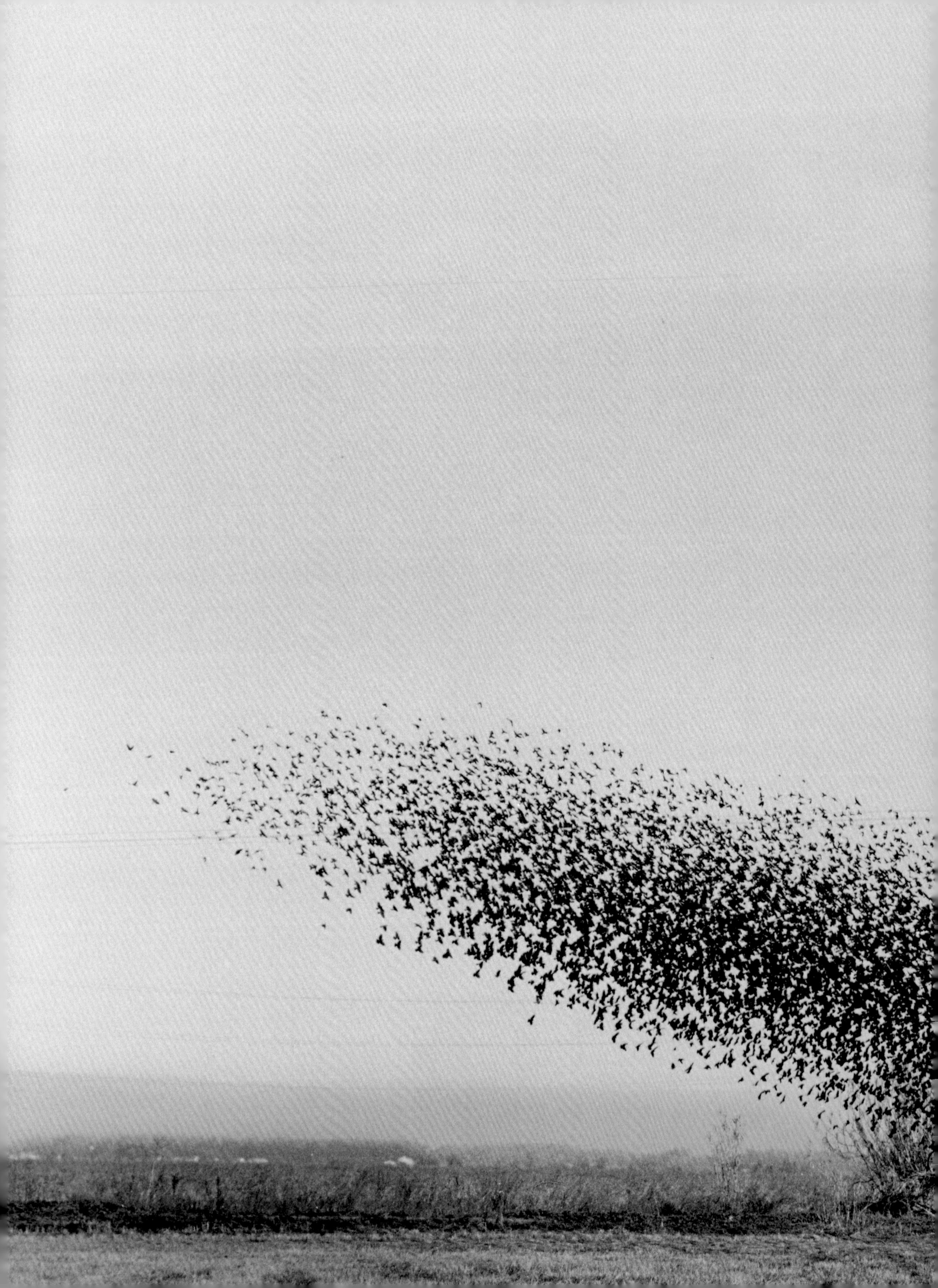

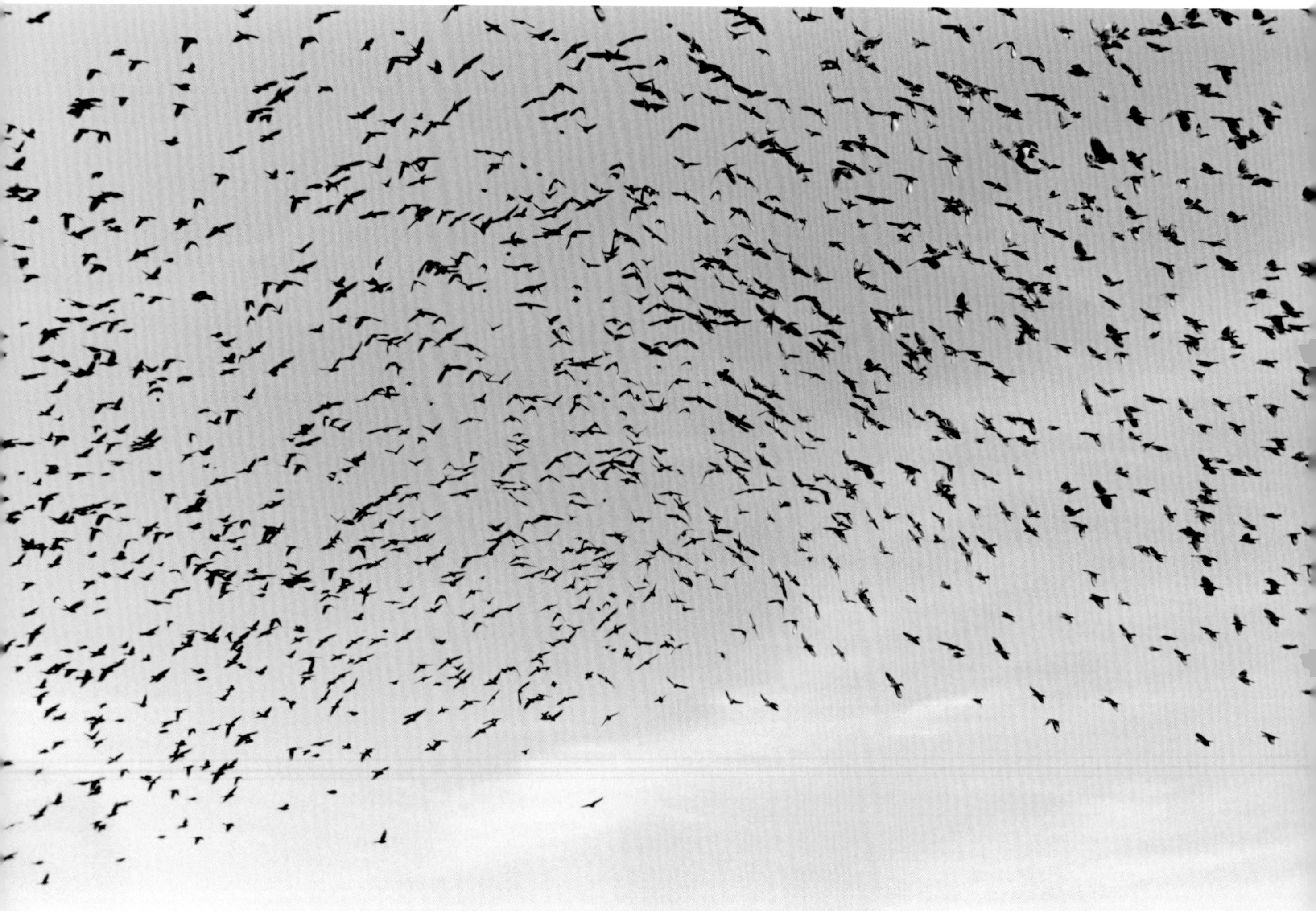

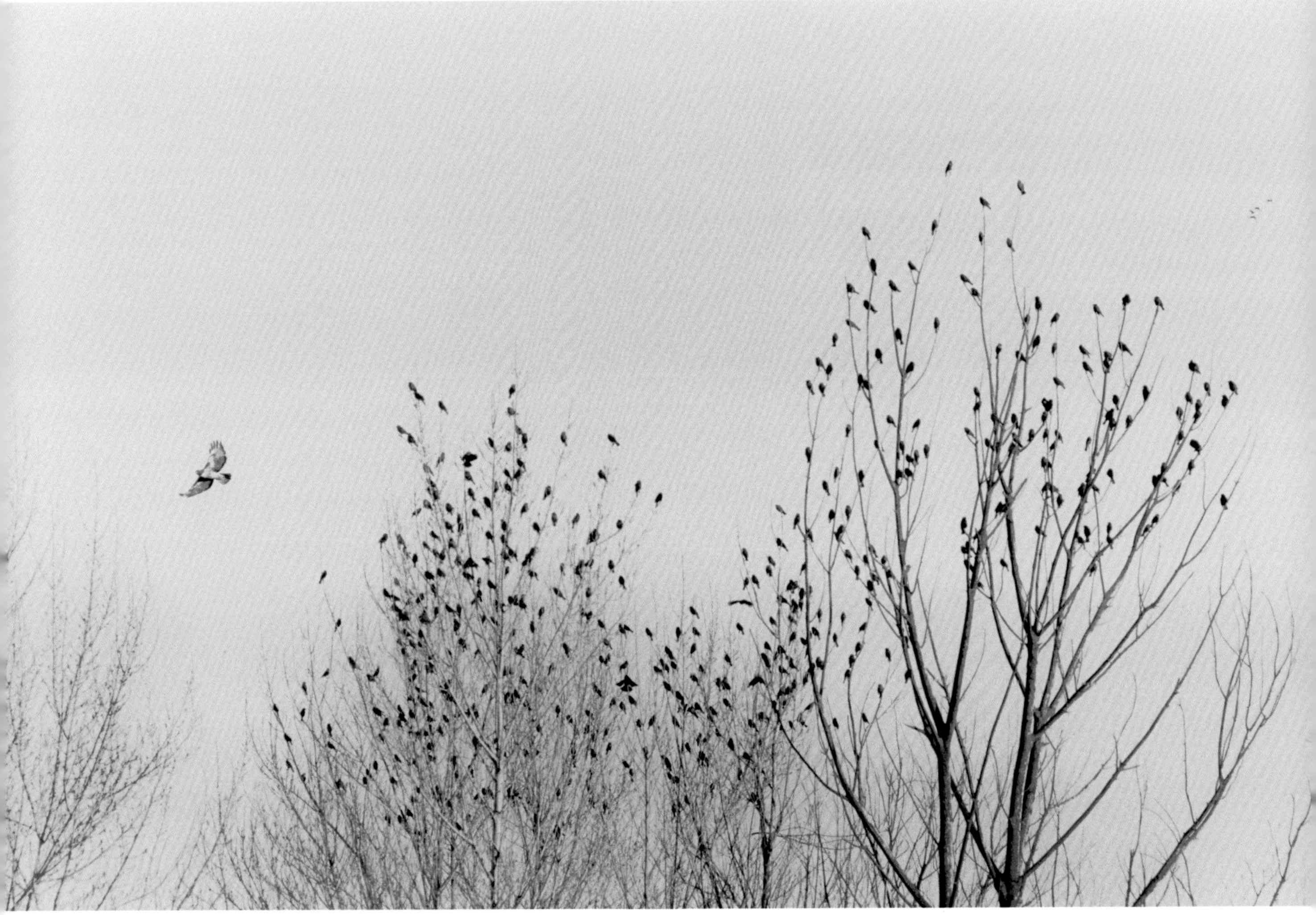

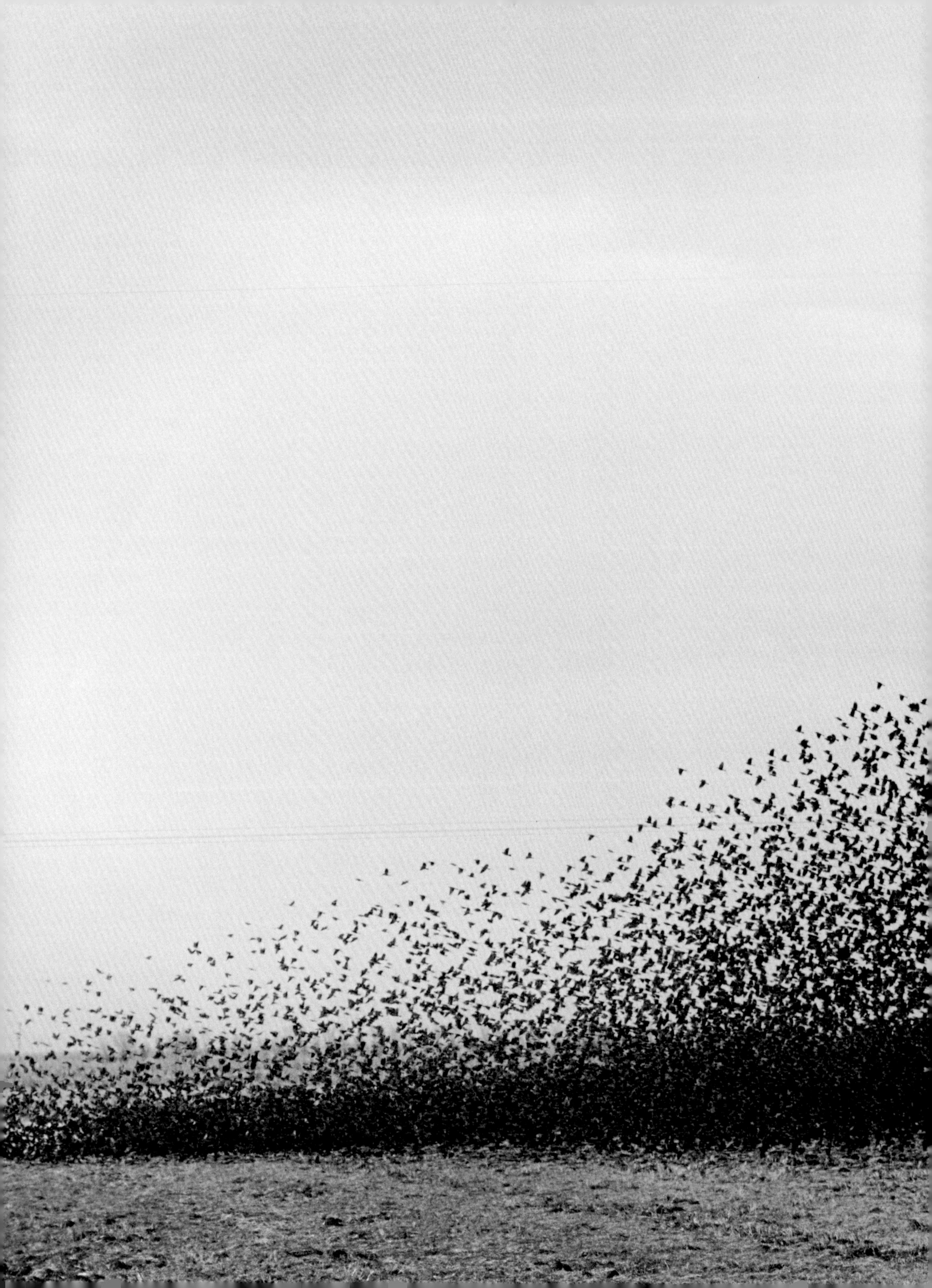

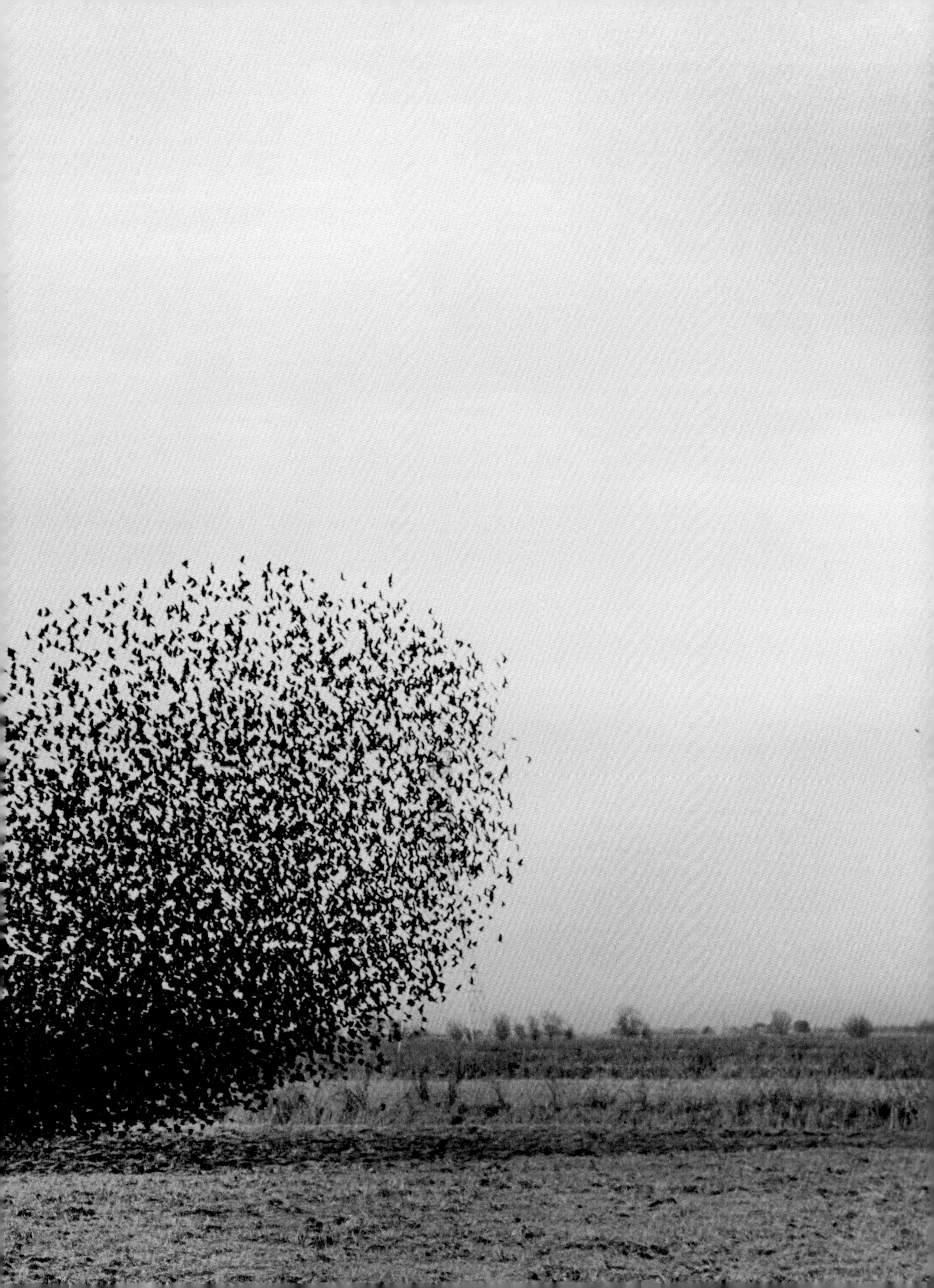

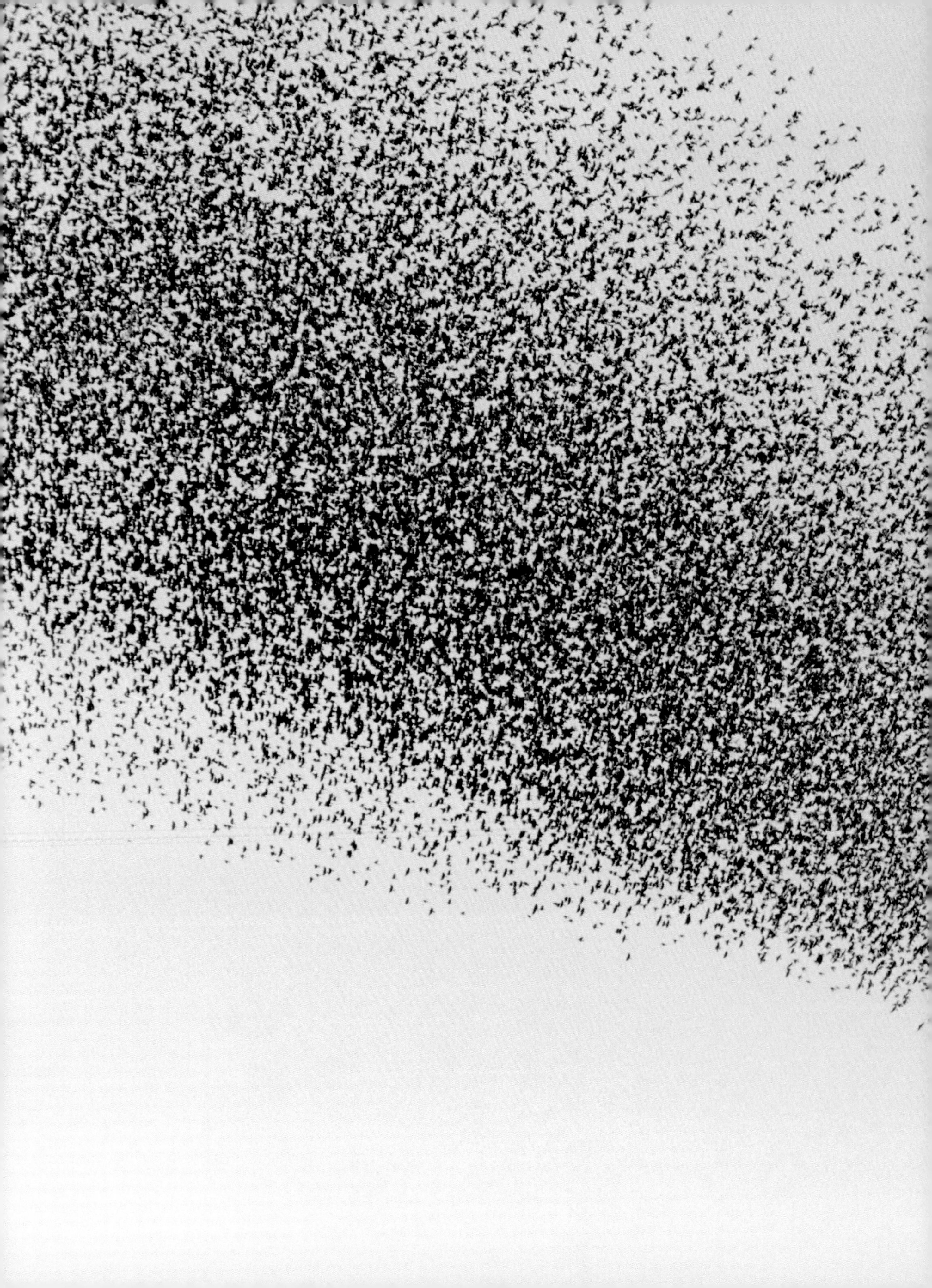

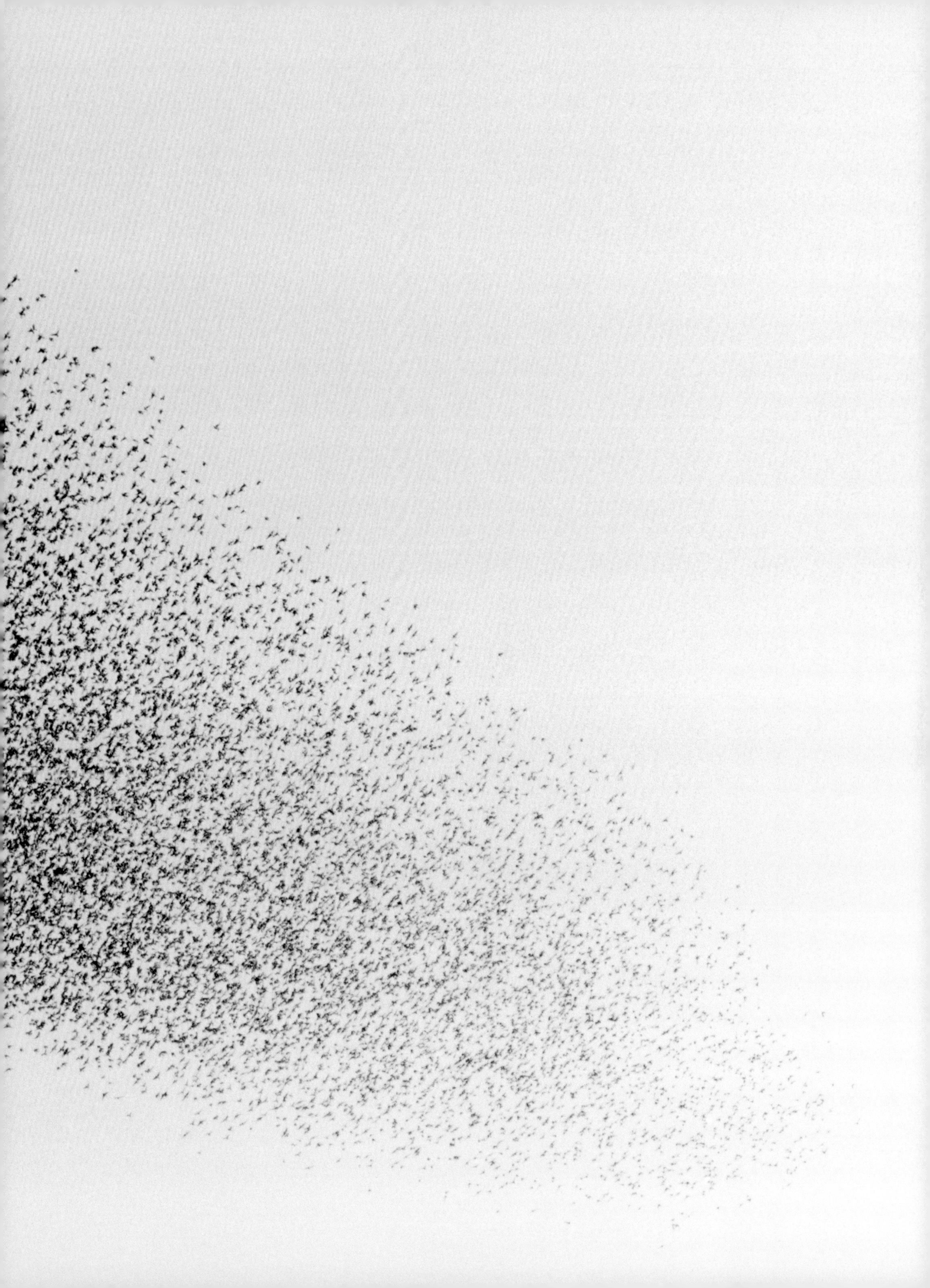

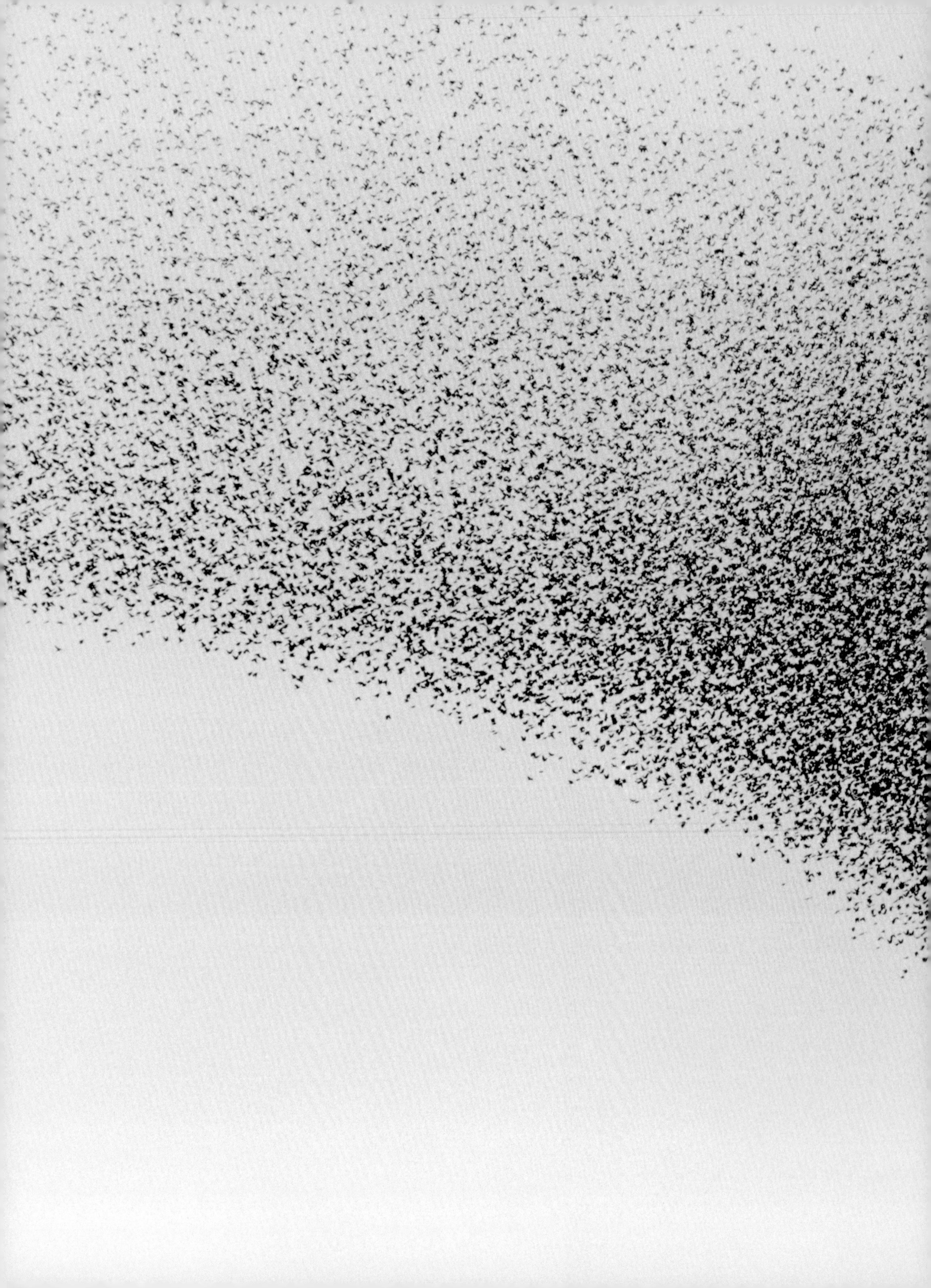

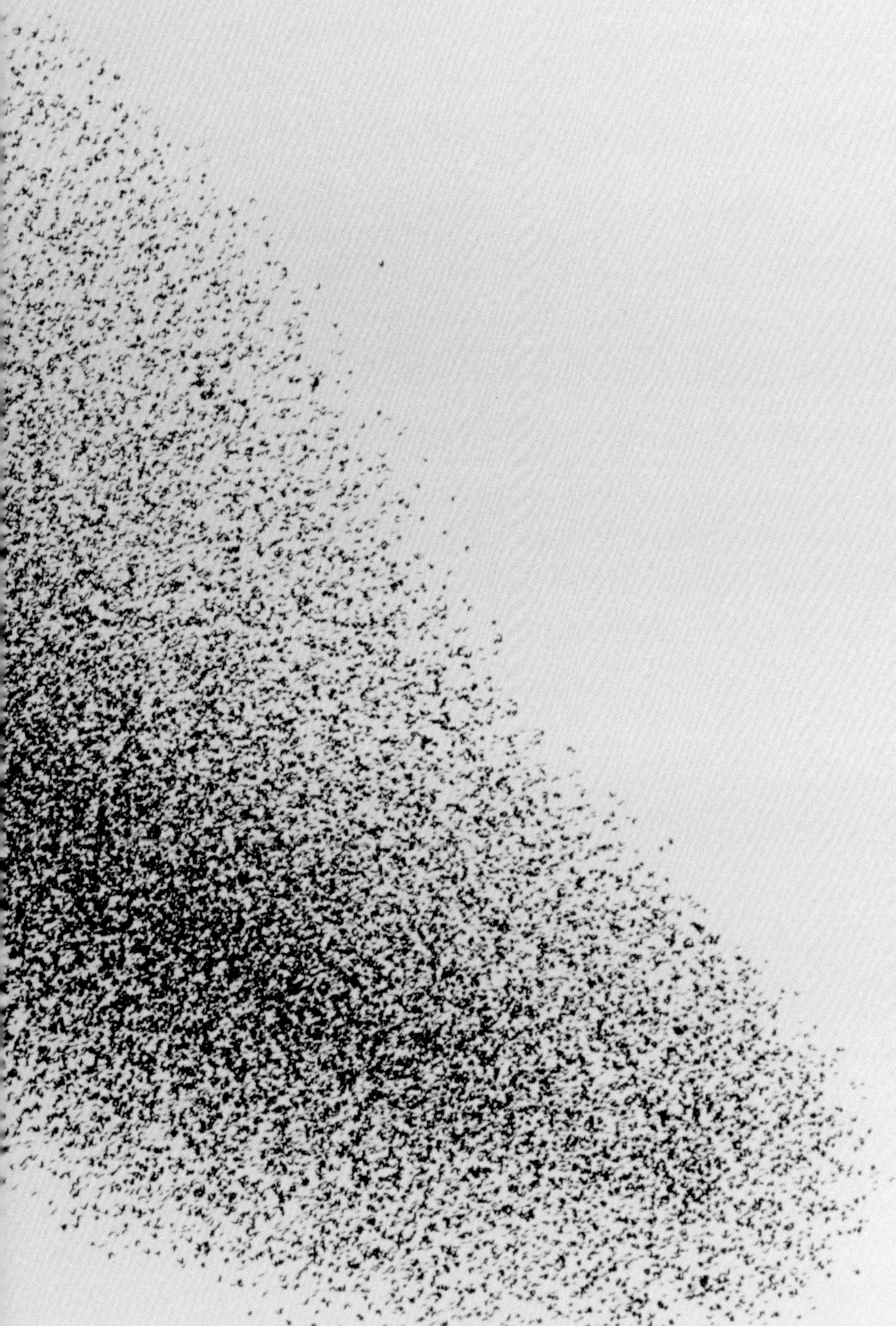

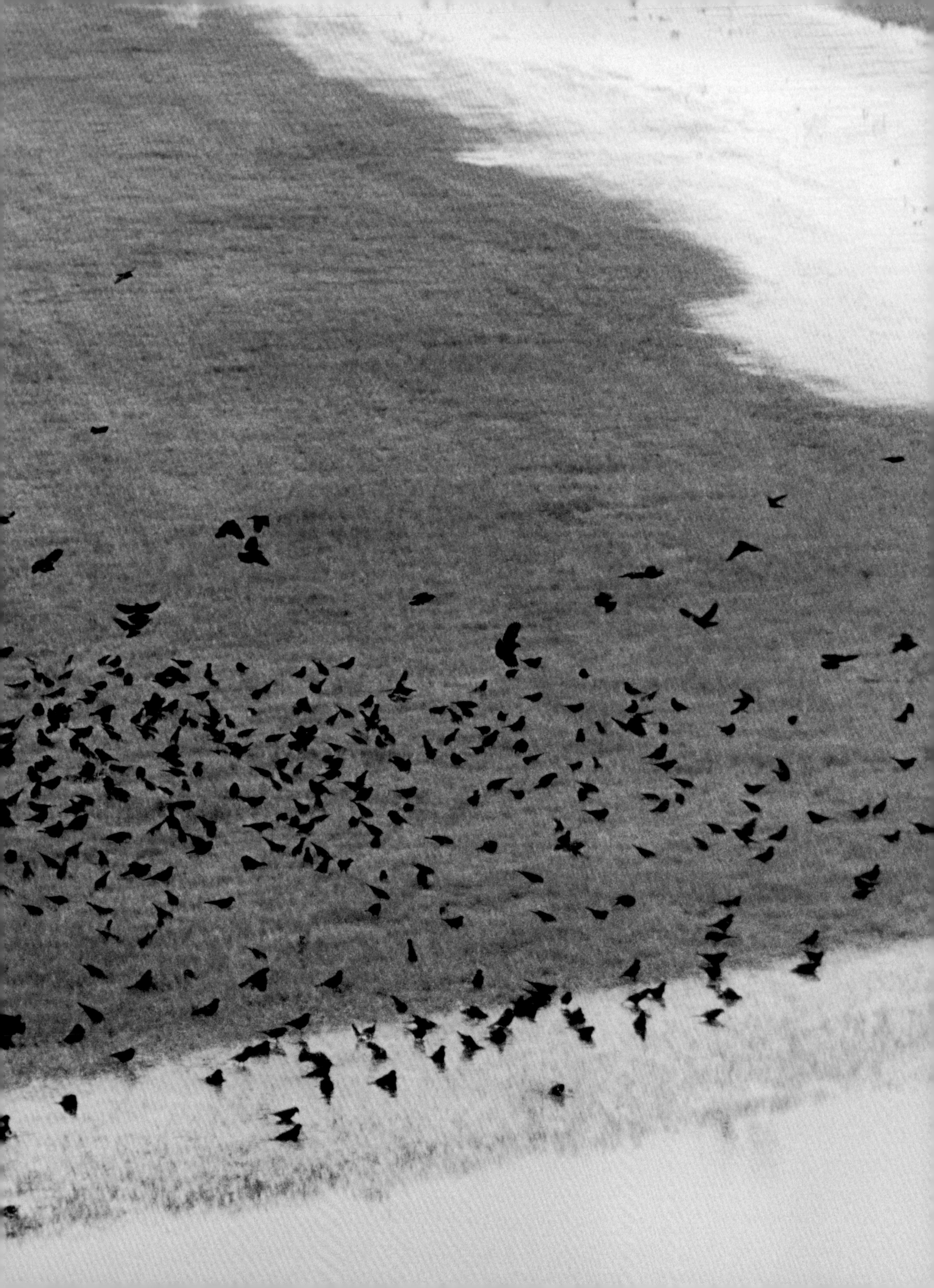

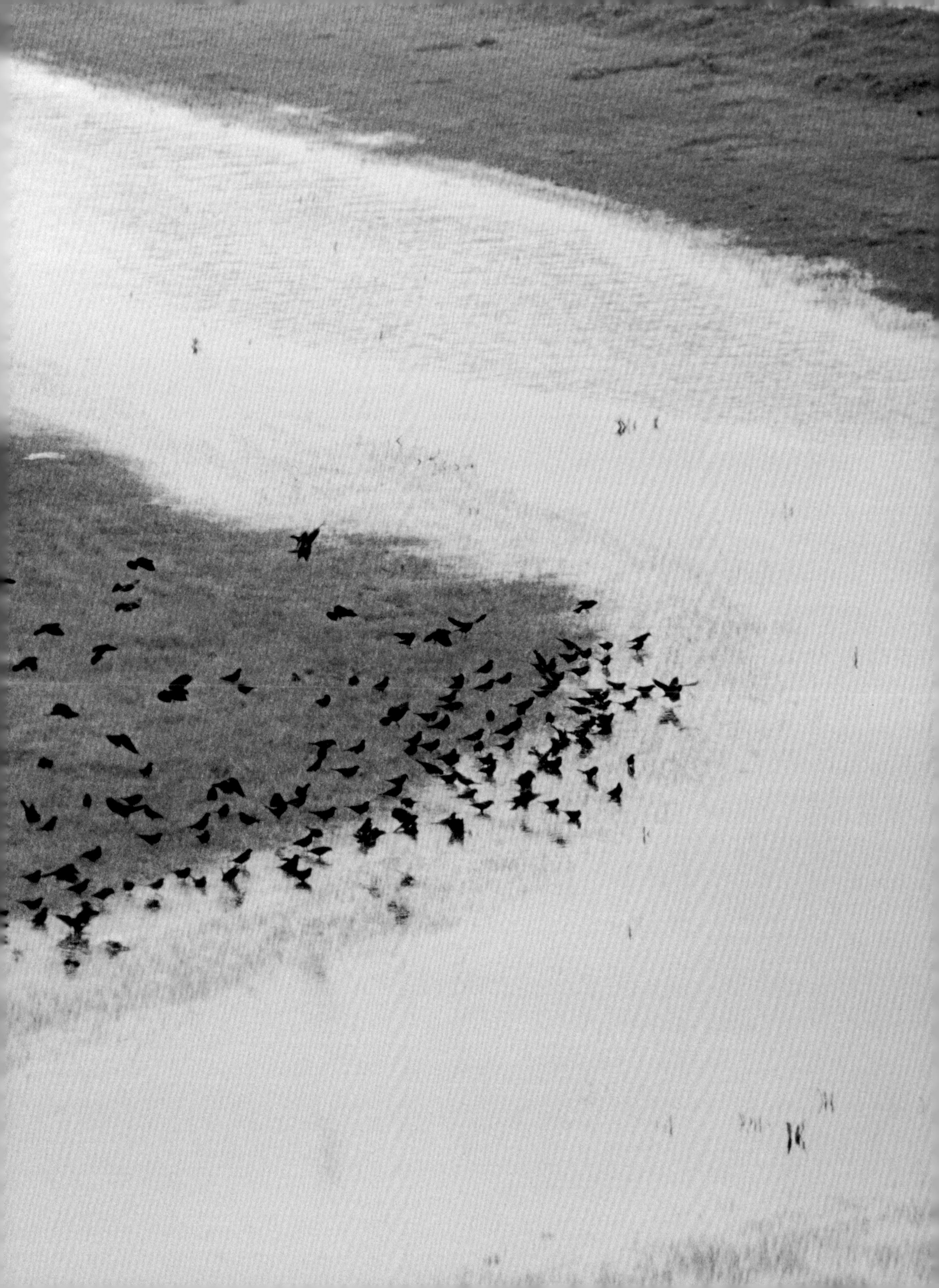

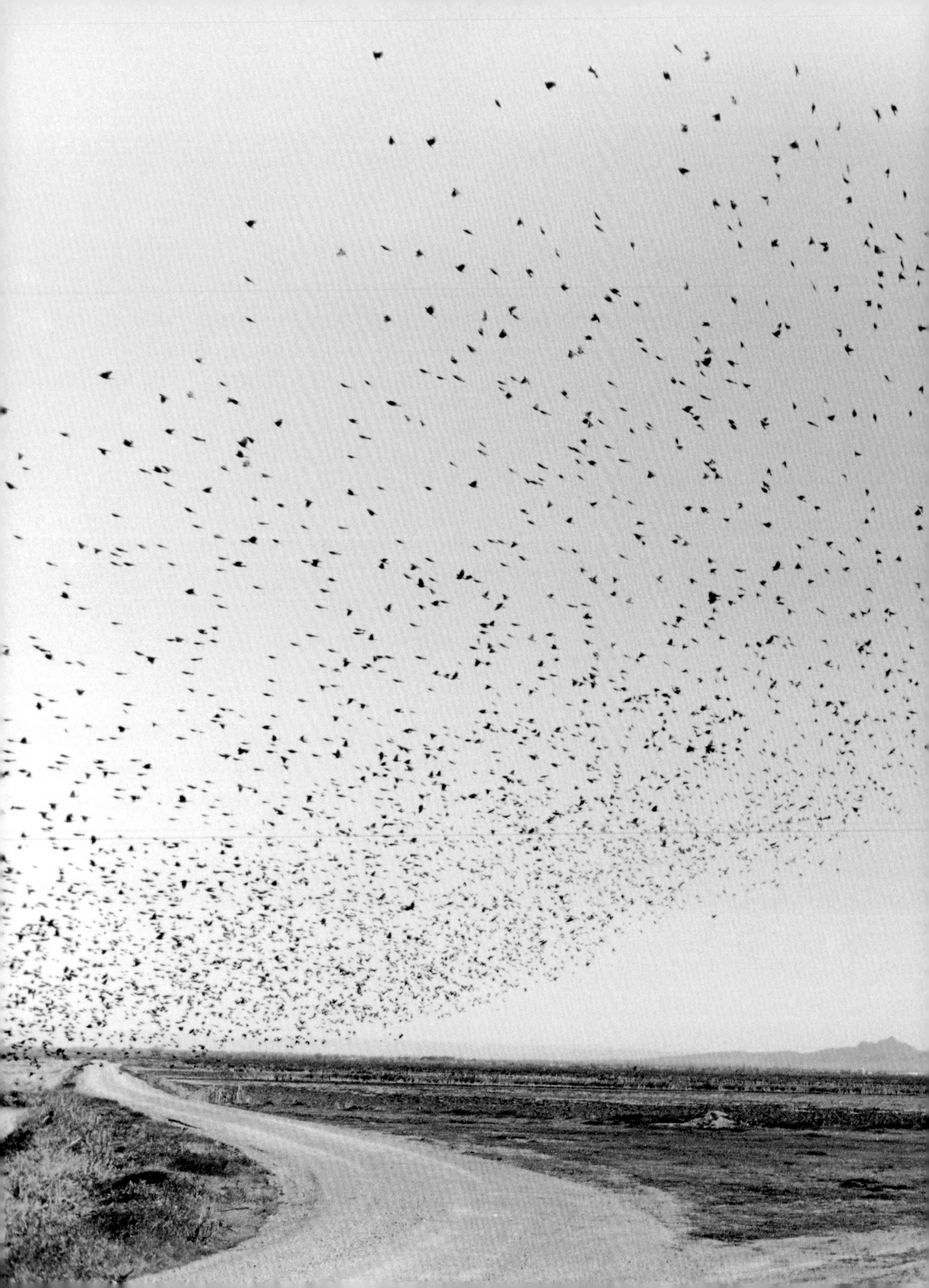

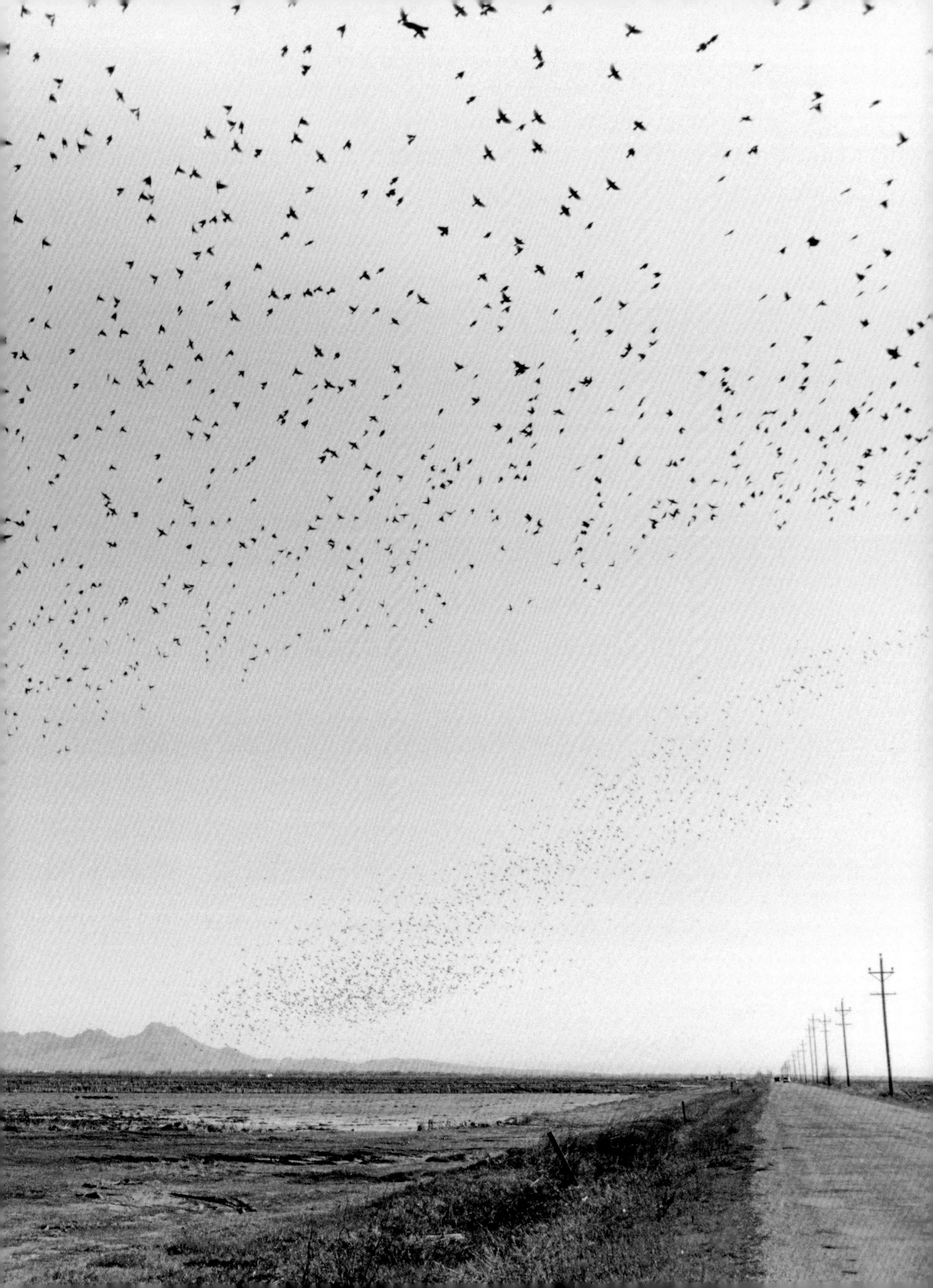

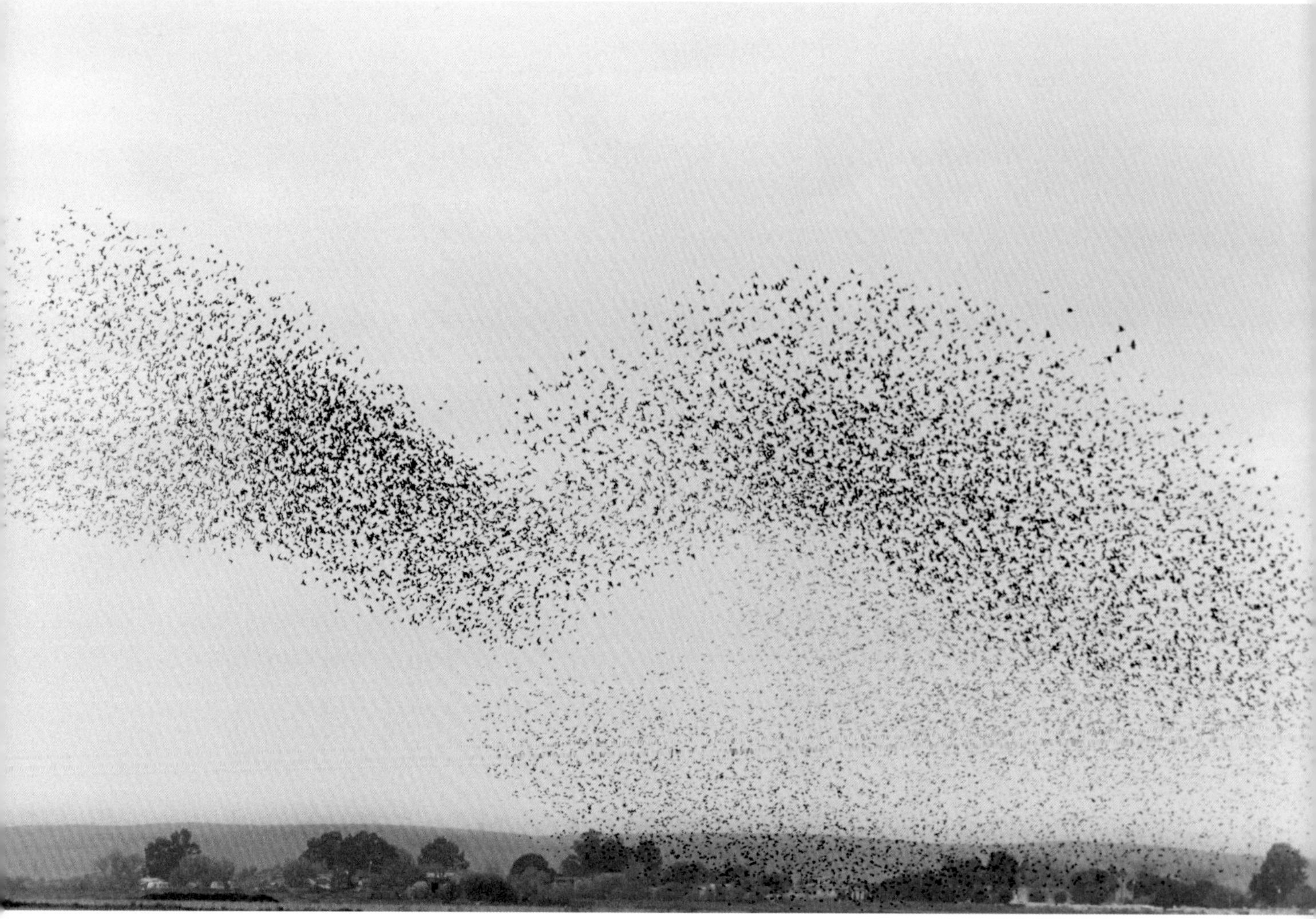

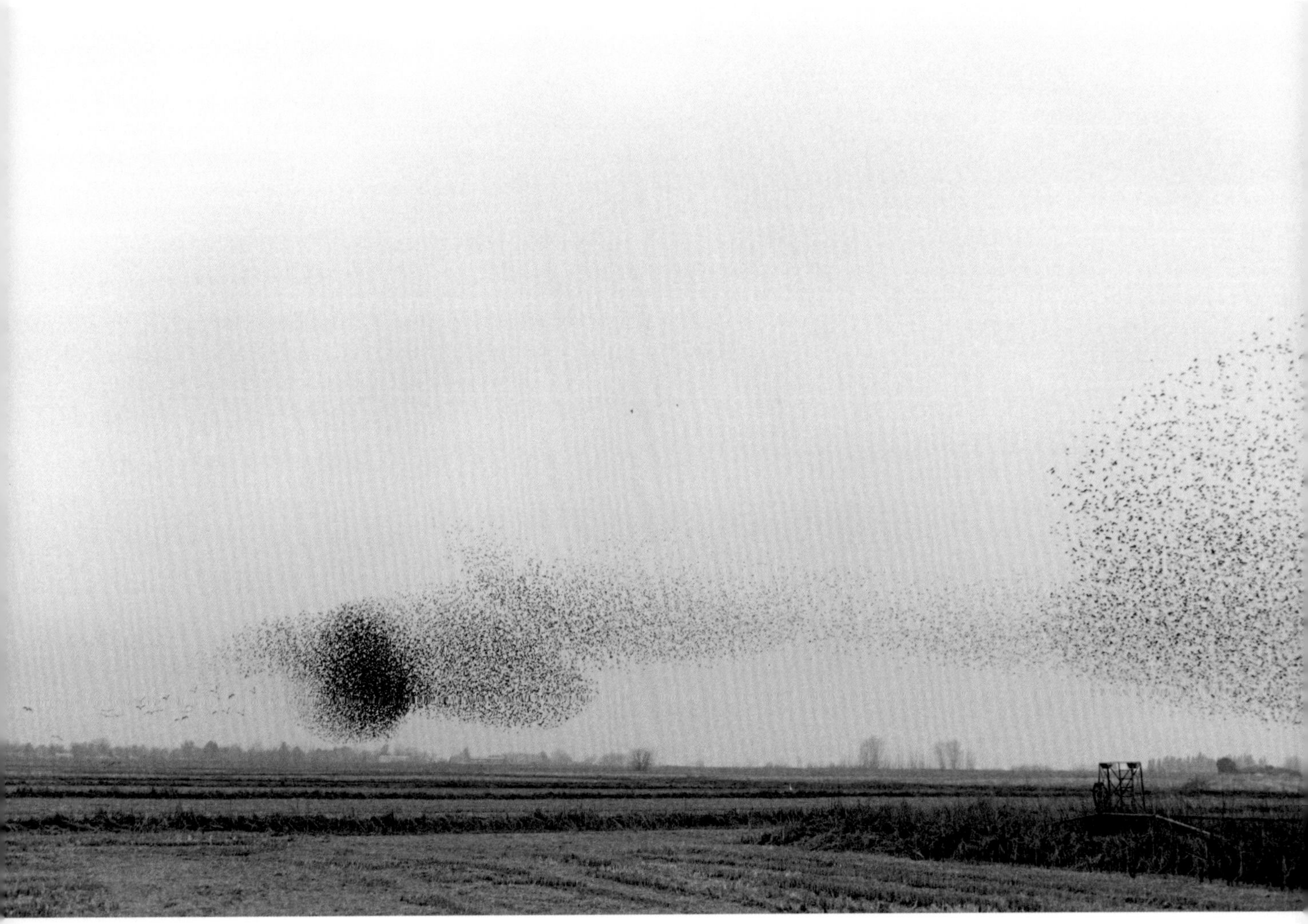

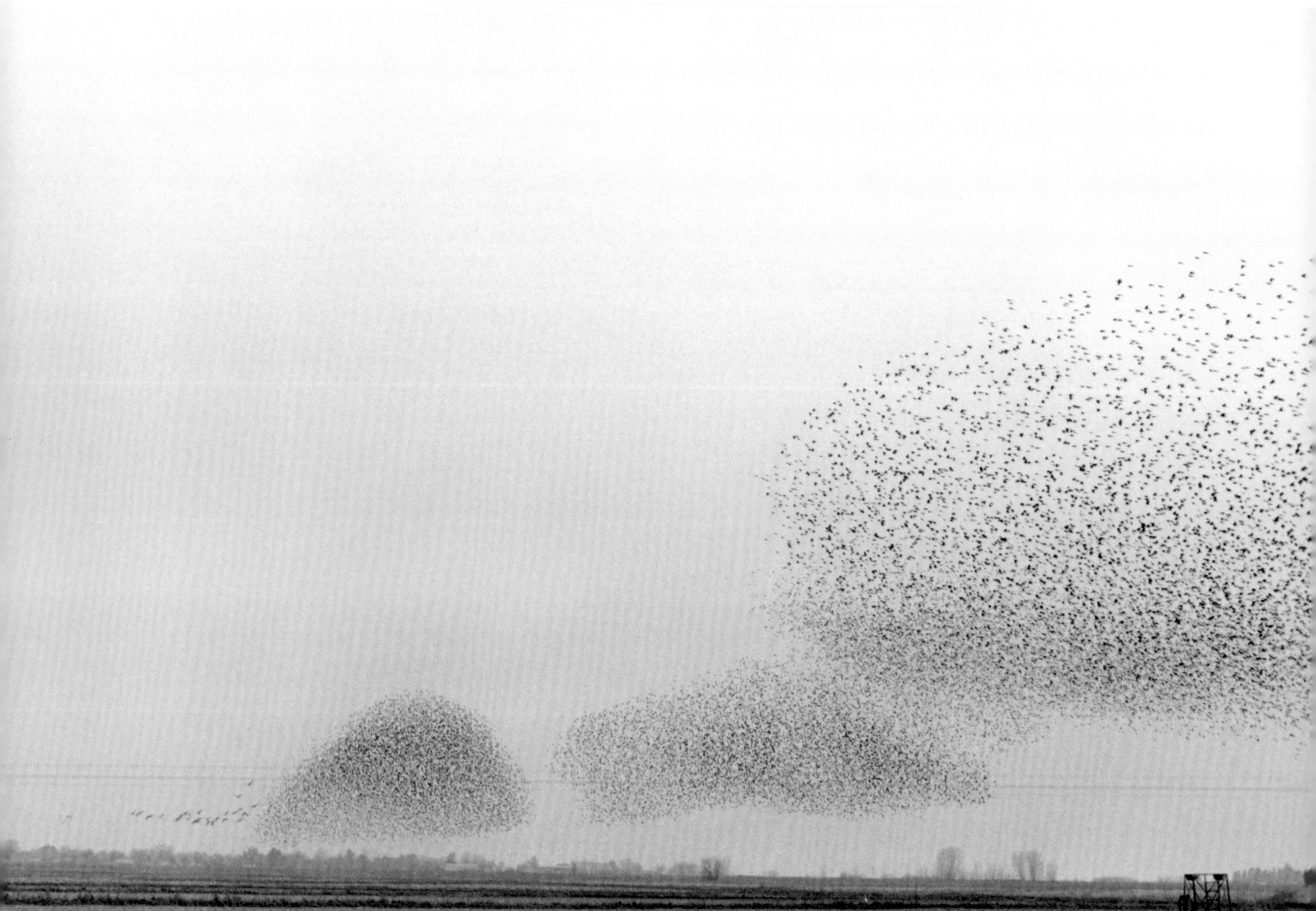

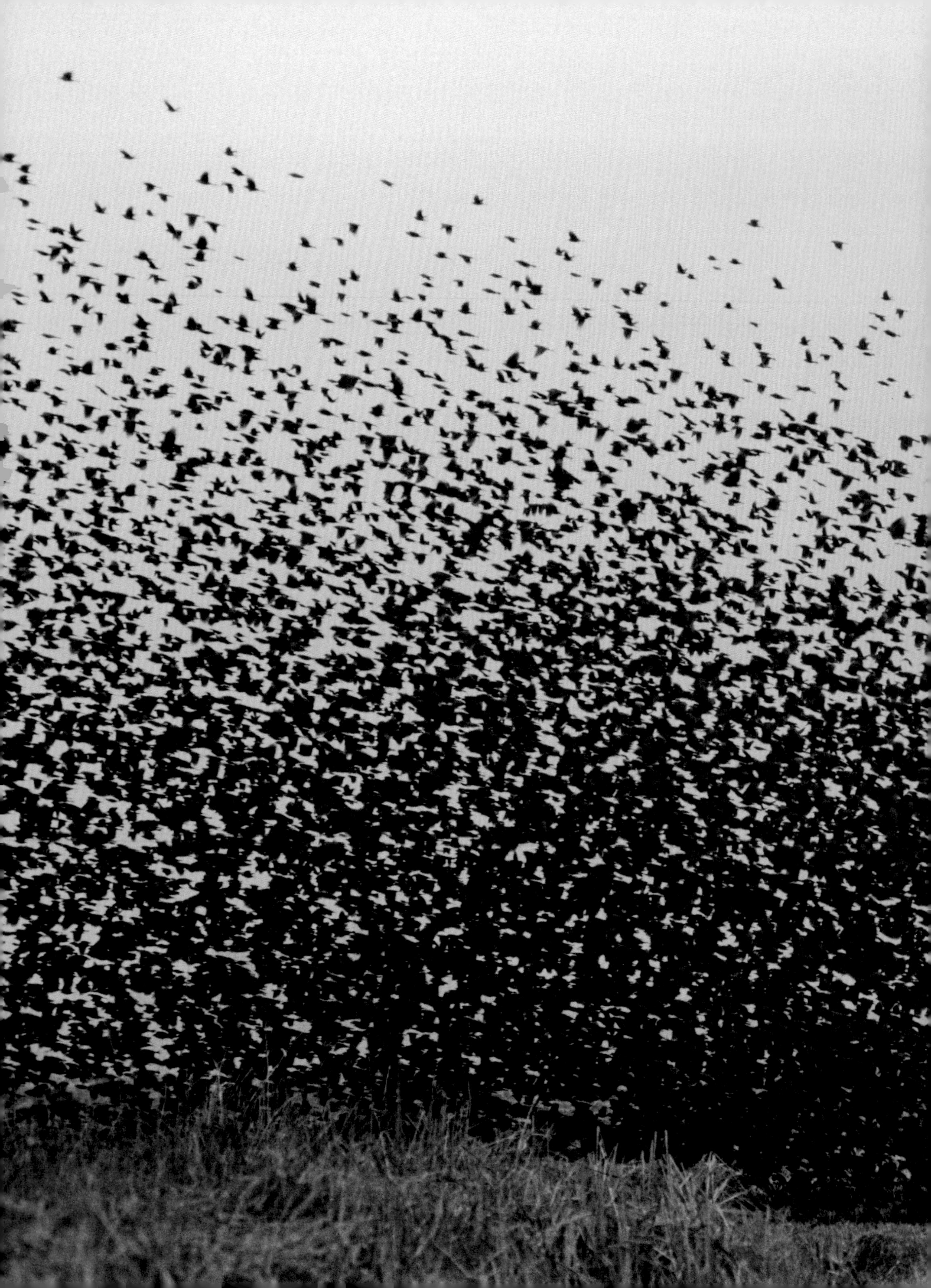

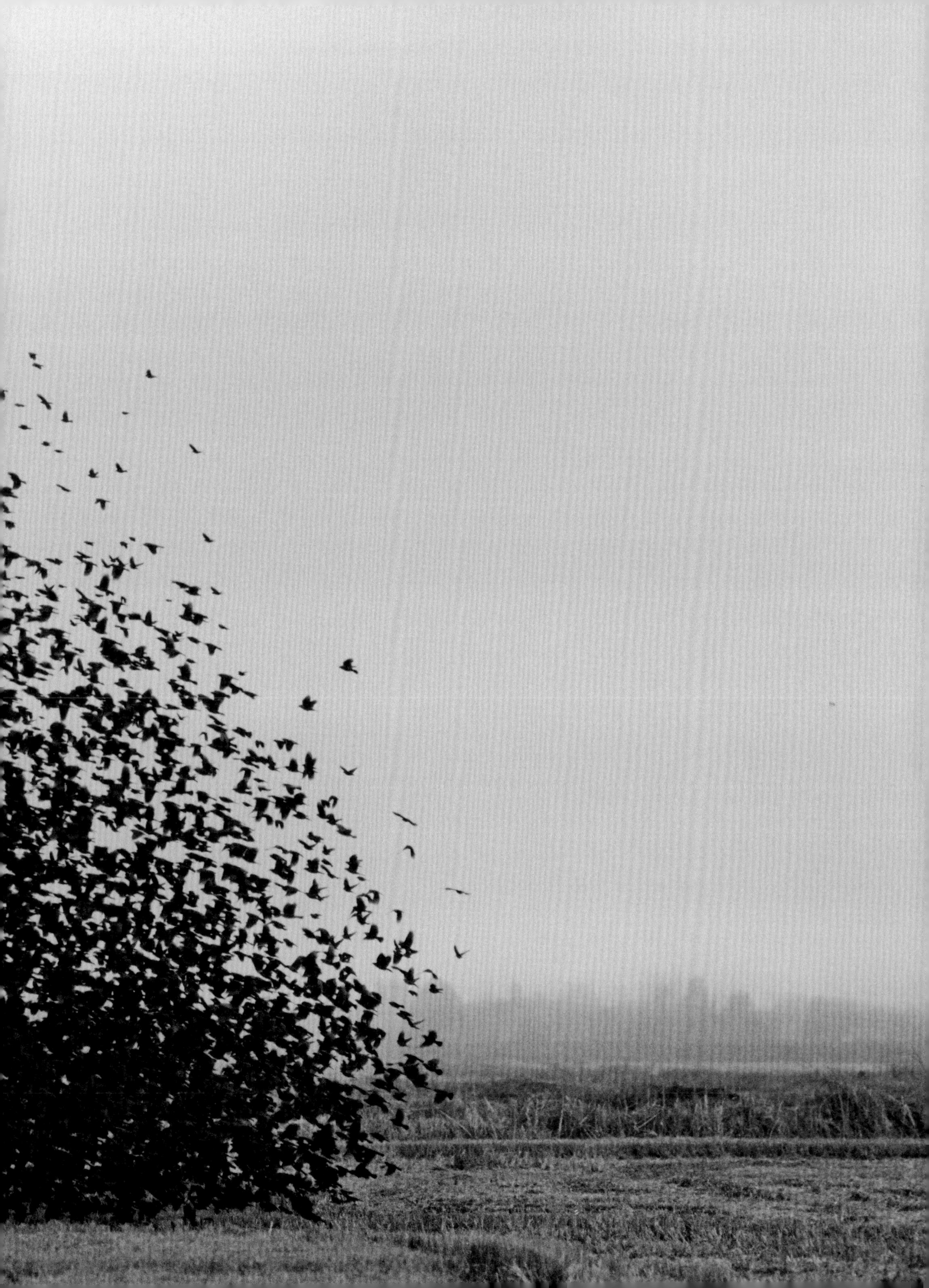

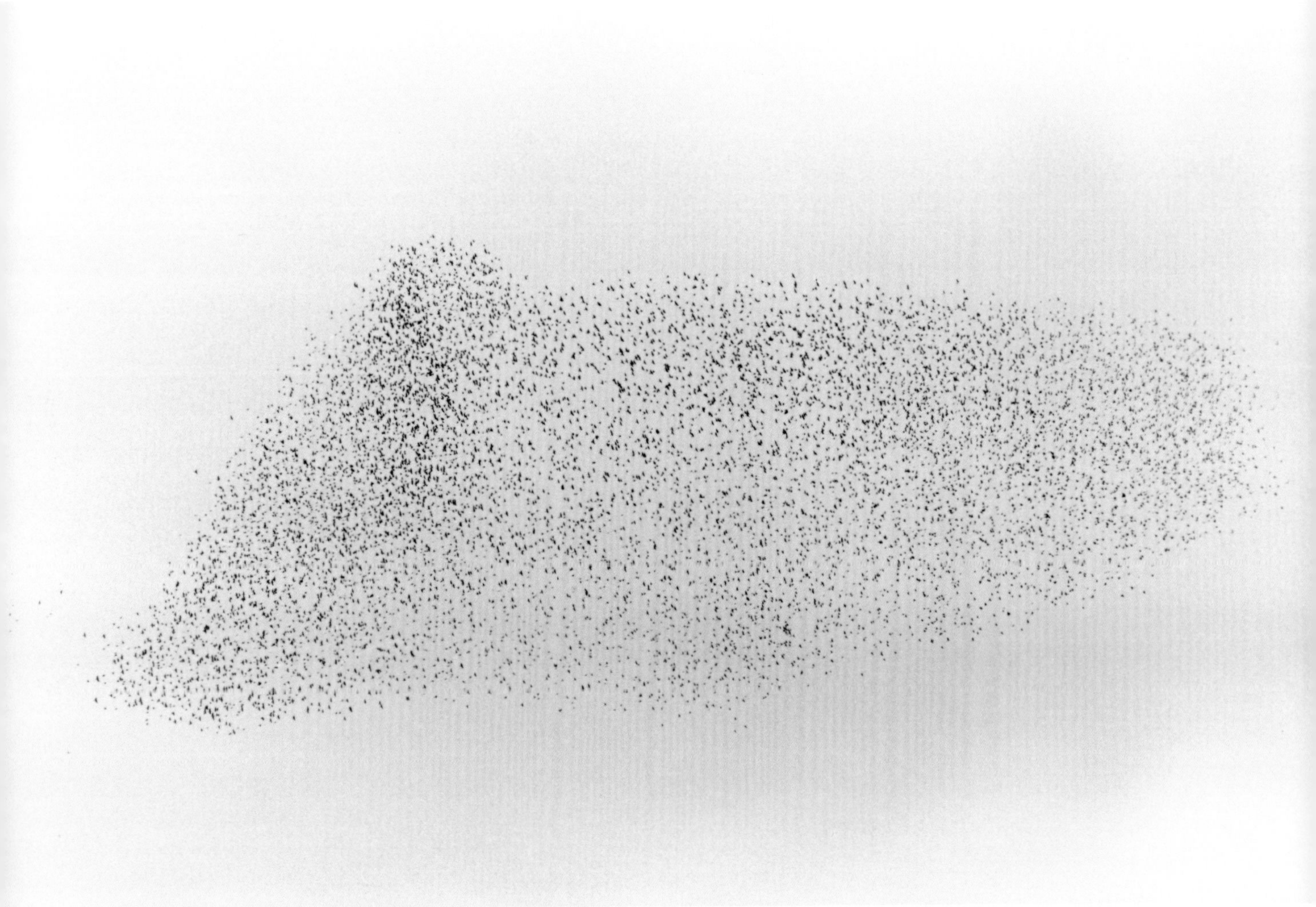

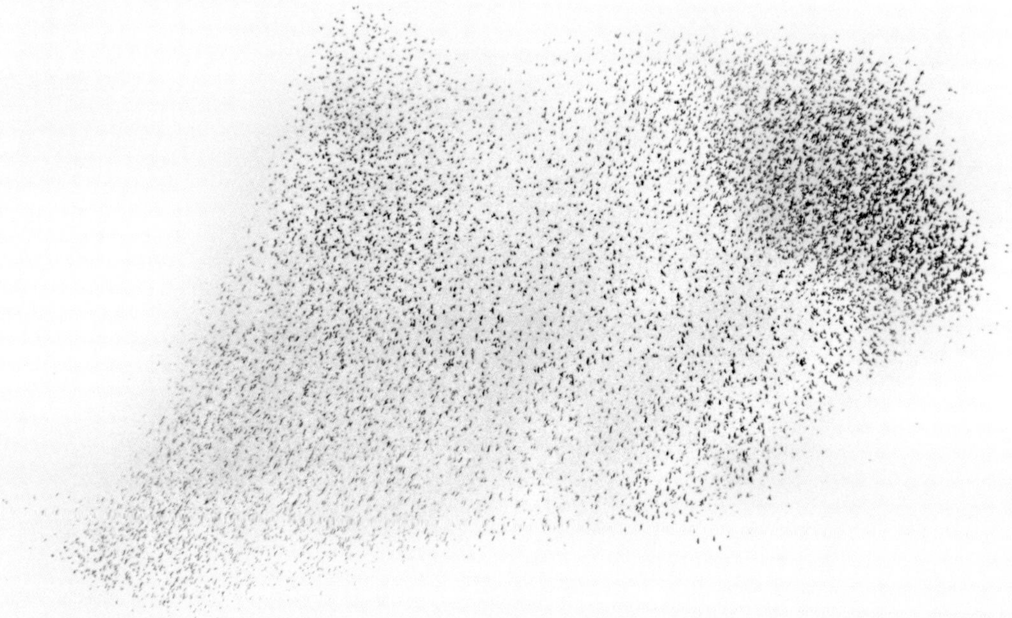

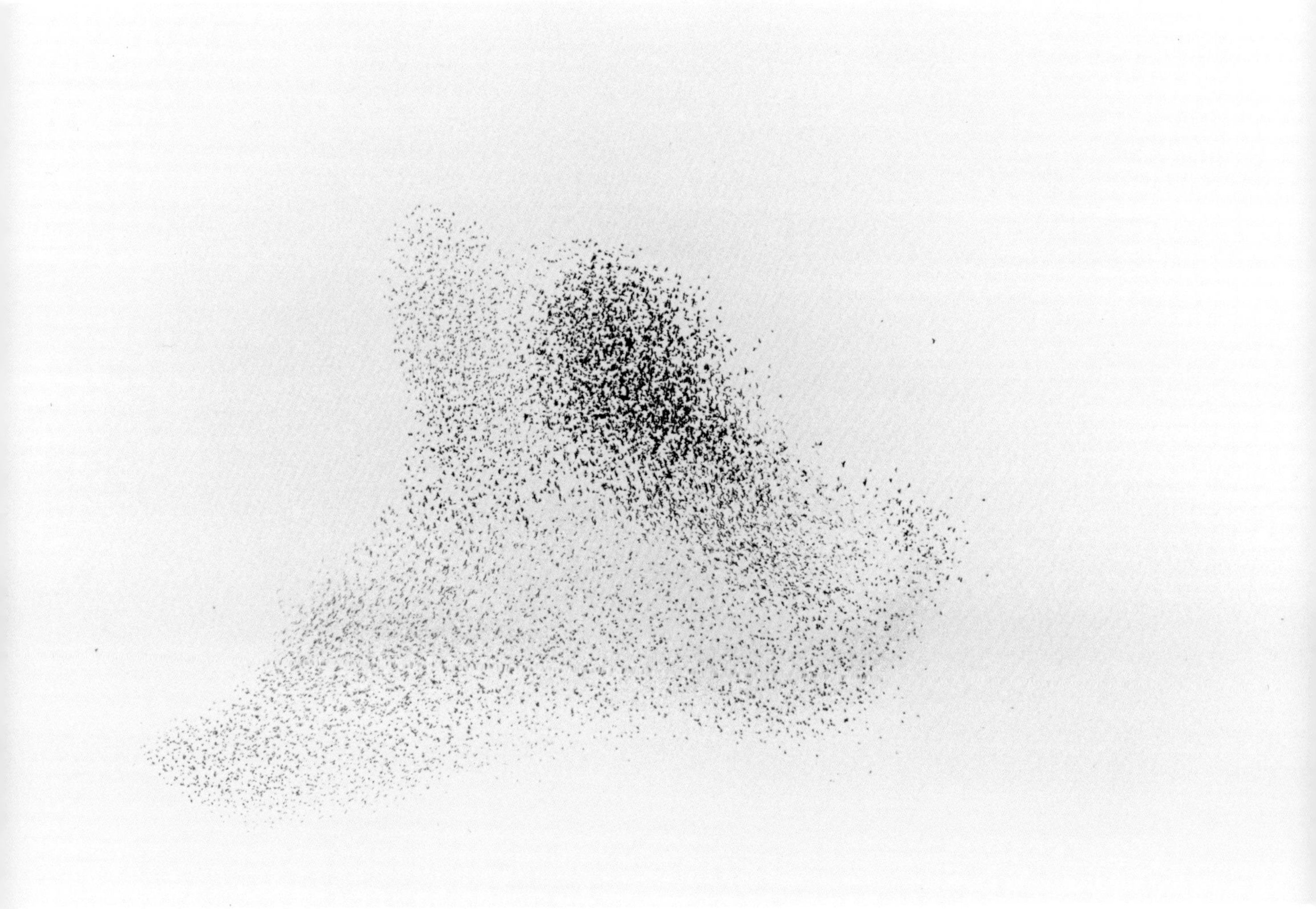

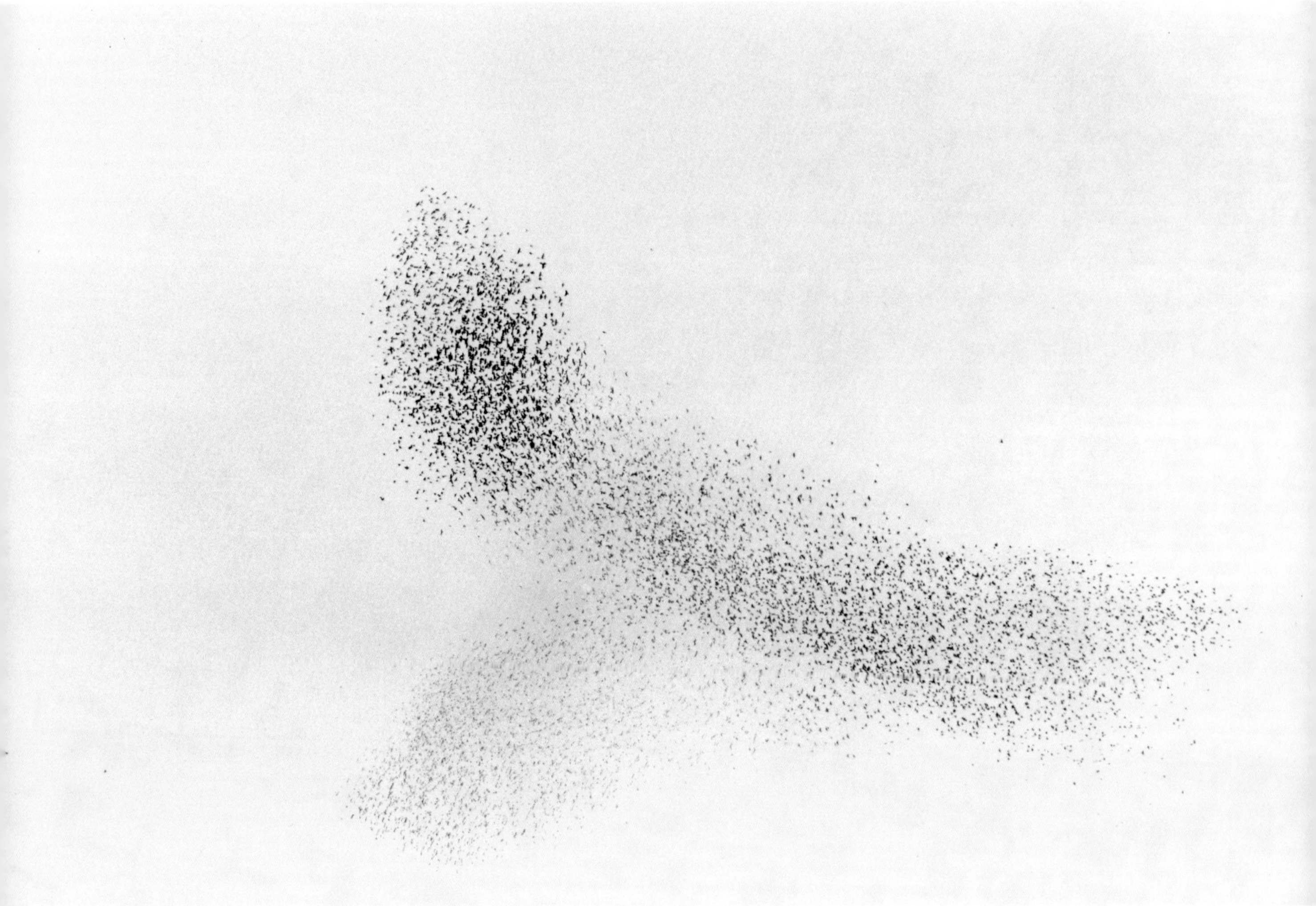

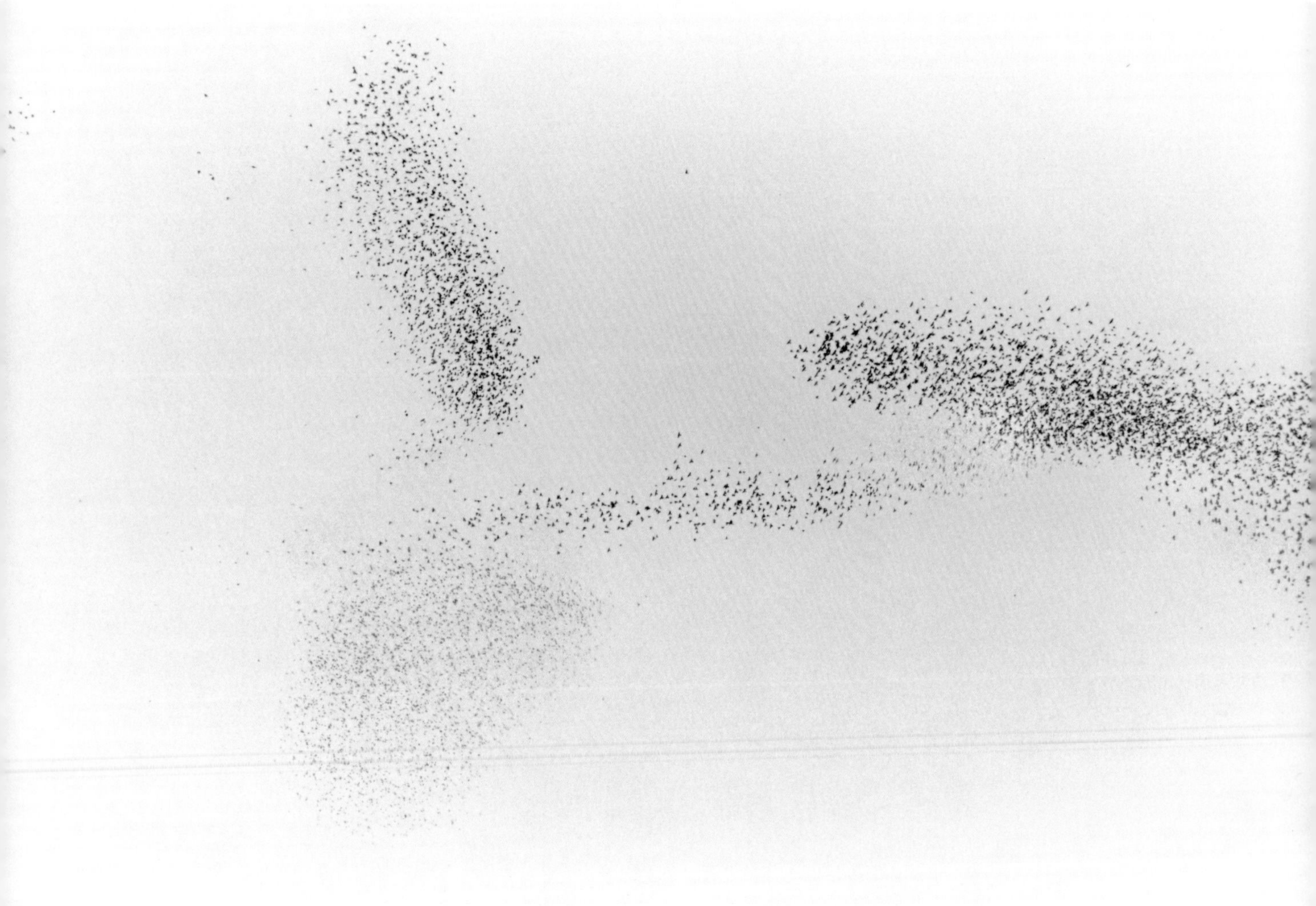

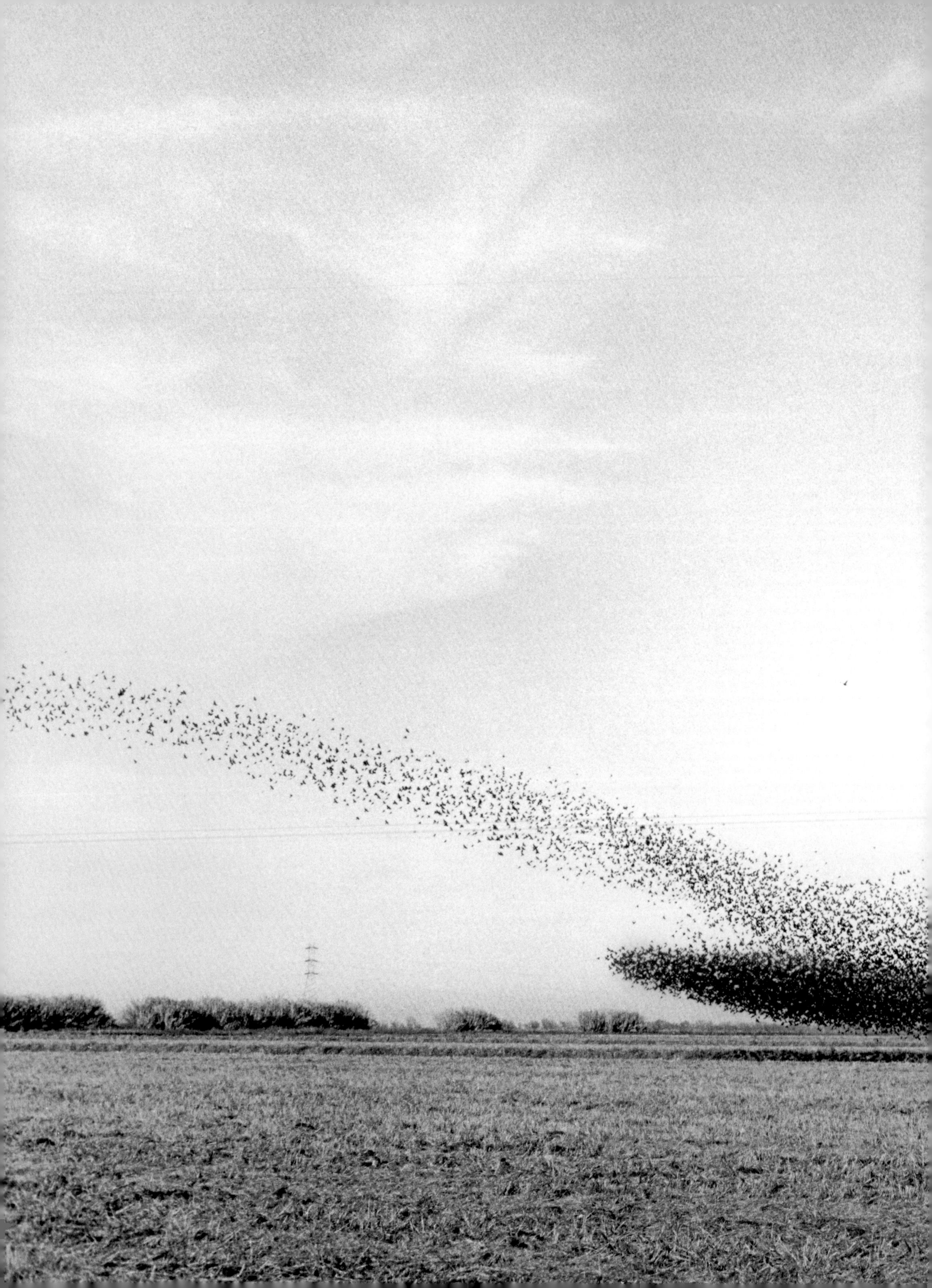

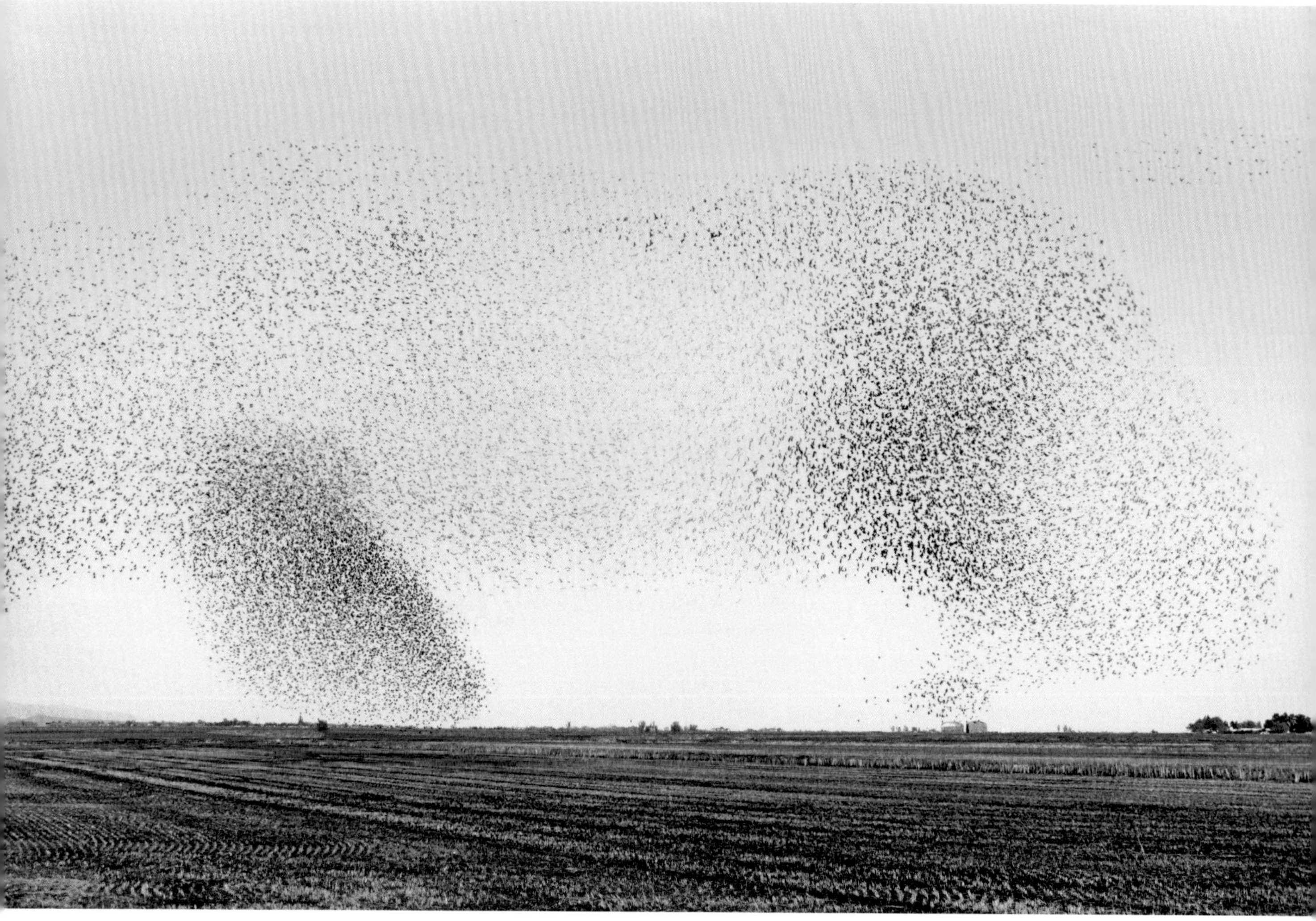

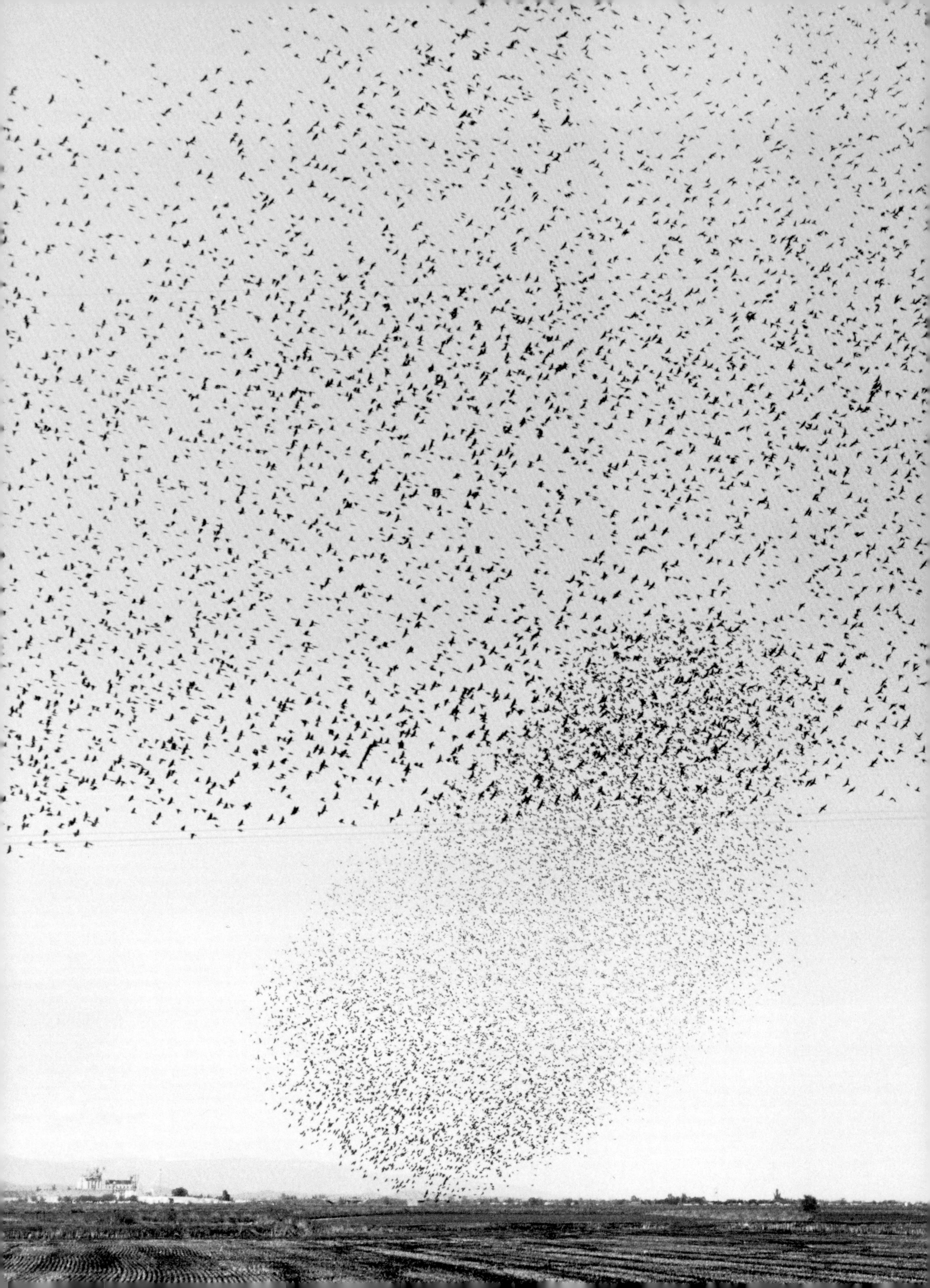

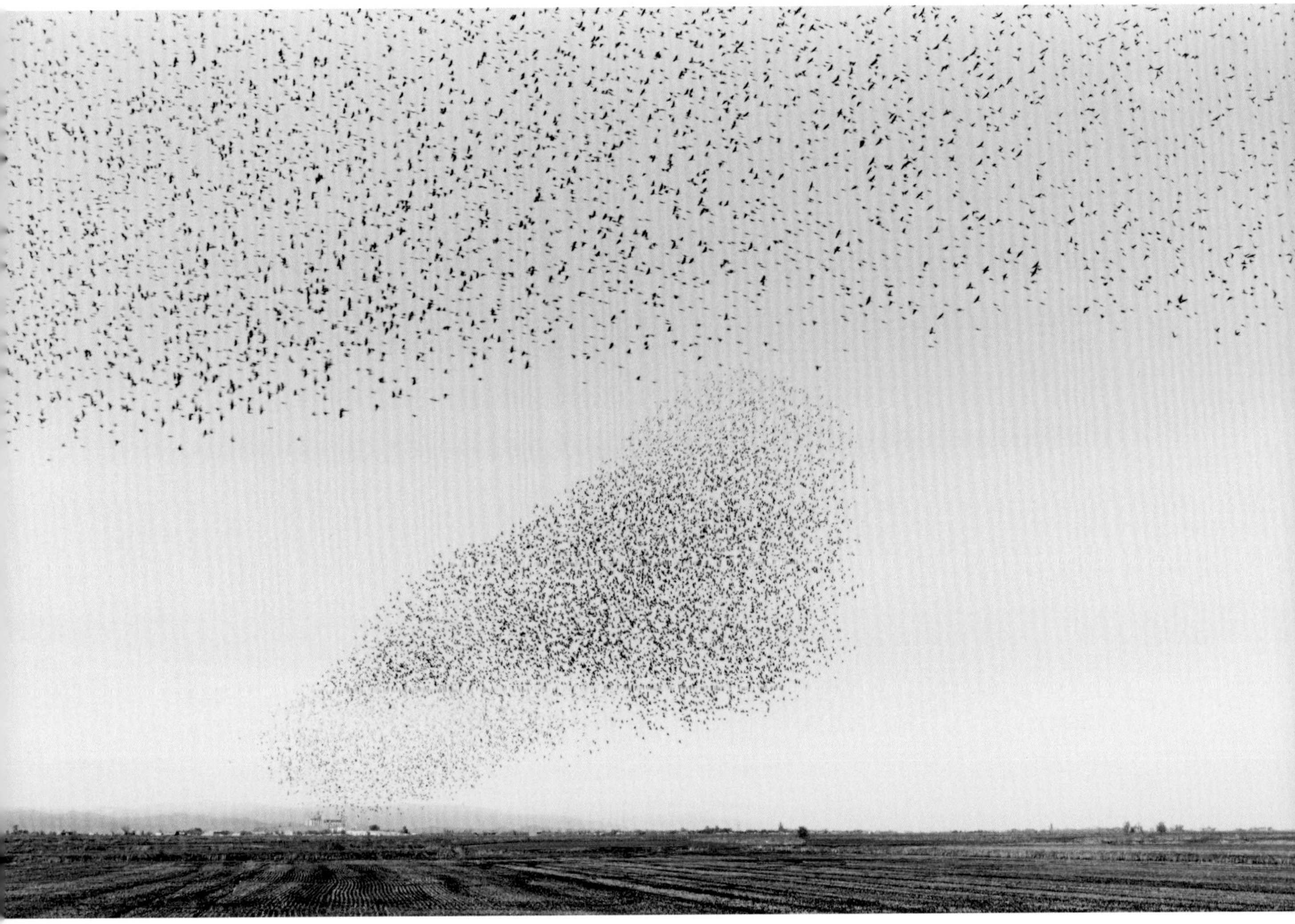

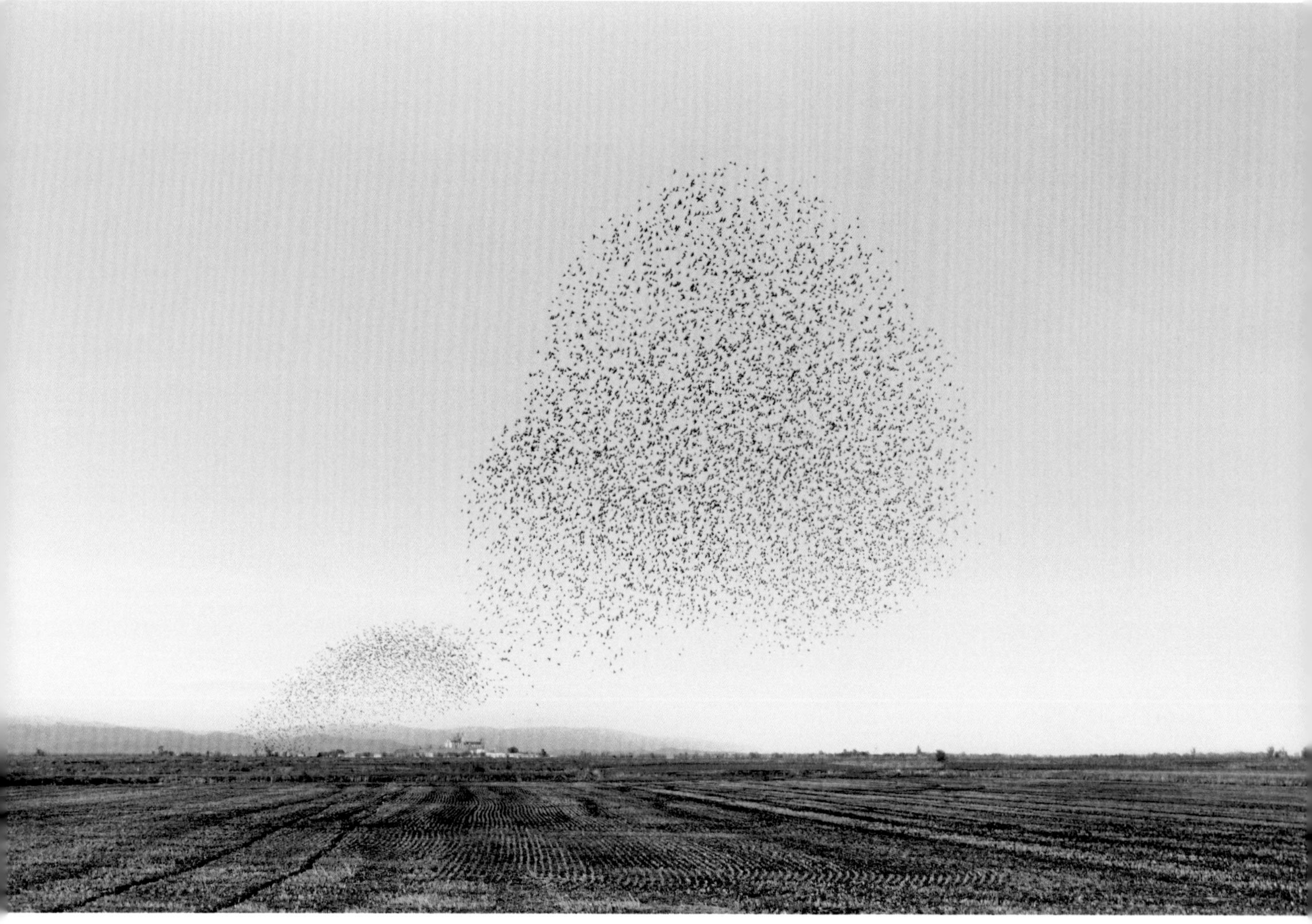

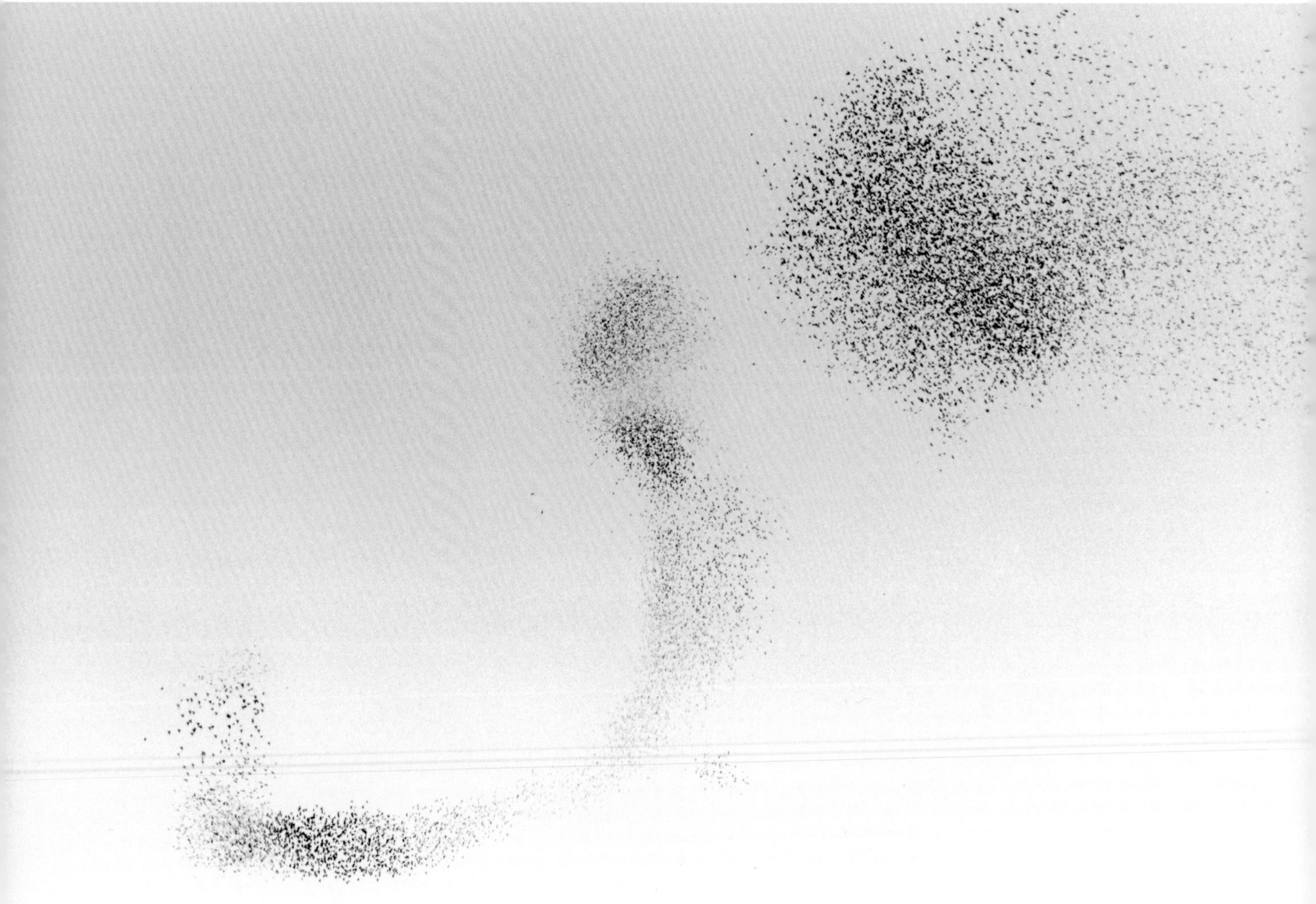

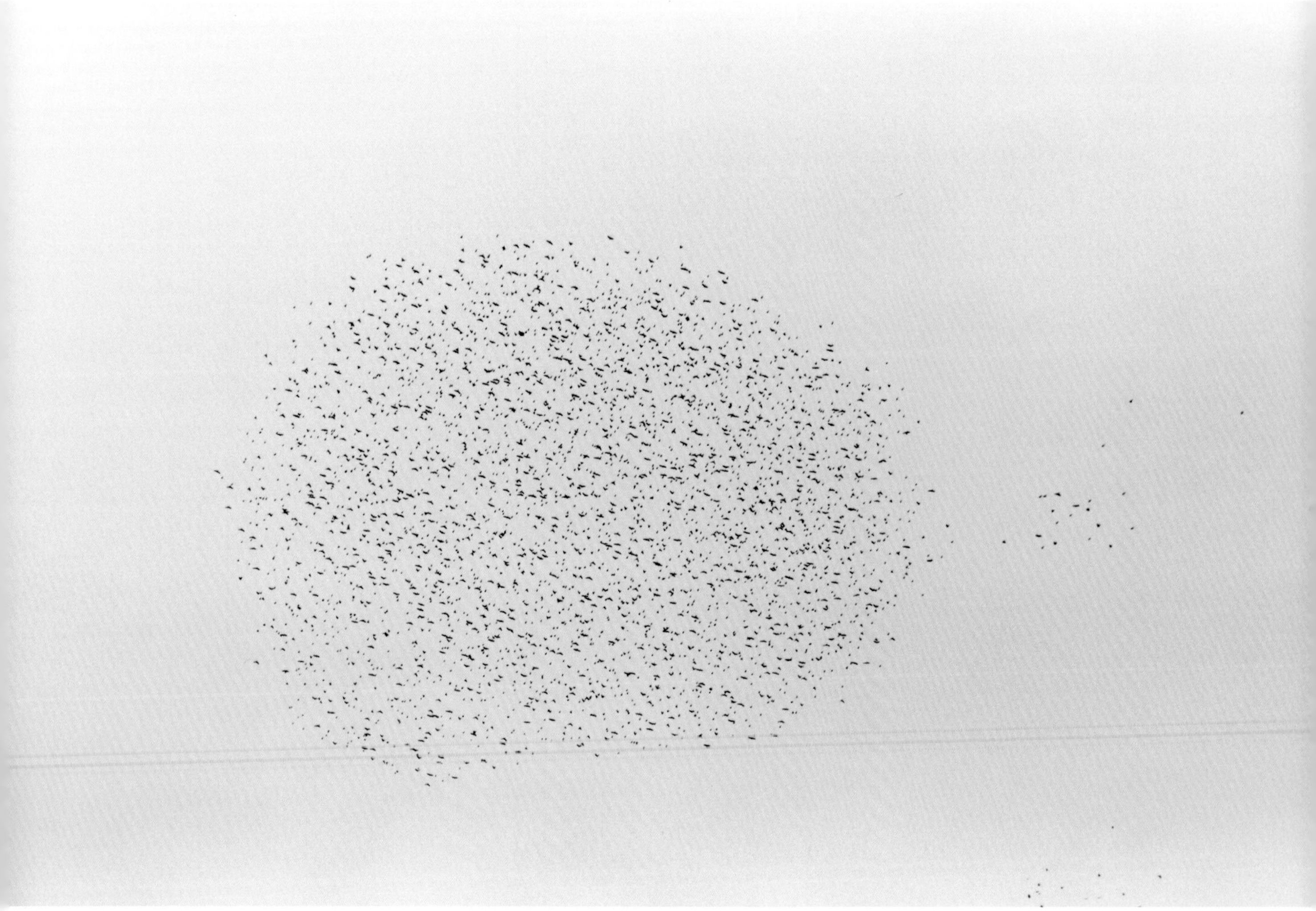

Control without Hierarchy

Deborah M. Gordon

Because most of the dynamic systems that we design, from machines to governments, are based on hierarchical control, it is difficult to imagine a system in which the parts use only local information and the whole thing directs itself. To explain how biological systems operate without central control—embryos, brains, and social insect colonies are familiar examples—we often fall back on metaphors from our own products, such as blueprints and programs. But these metaphors don't correspond to the way a living system works, with parts linked in regulatory networks that respond to environment and content.

A flock of birds turning in the sky is doing something that people don't know how to do: moving together, beautifully, without a leader or choreographer. It's a spectacular version of the collective behavior that goes on everywhere, in groups of animals and among cells within our bodies. Each bird responds to its neighbors, and the whole flock turns. Birds interact by watching each other. Each anchovy in the school responds to the currents made by neighboring fish, and the school shoots away from a shark, or splits around an obstacle, to form again on the other side. Termite colonies build enormous mounds held up by a system of arches. An arch is built without a plan: two groups of workers each build a pillar, then pile soil on top of each pillar, leaning towards the smell of their nest mates, so that the two pillars bend and meet to form an arch. On a smaller scale, the pattern of neurons firing off and on allows your brain to think, and the response of cells to neighboring ones leads embryos to develop muscle and bone out of a cluster of identical cells.

I study ant colonies, and how they get things done without any central control. The contacts among ants are messages without content. What matters is merely the fact that certain ants meet. They don't tell each other anything. The pattern of interactions is the message. What each ant does depends on the rate at which it meets other ants. The rate of encounter each ant experiences with ants performing a certain task depends on how many ants are doing that task, and how quickly they get it done. Changing conditions change the encounter rate, and this creates feedback. In this way the colony adjusts to current conditions.

Life in all its forms is messy, surprising, and complicated. It's difficult to imagine how any social group could be organized without any hierarchy. We are used to hierarchy as the principle that organizes human institutions. Think of companies, armies, governments, orchestras, schools, and clubs—without any person directing another, or having more power than another. Although we are so accustomed to hierarchy that we think of it as necessary, it is rare in nature.

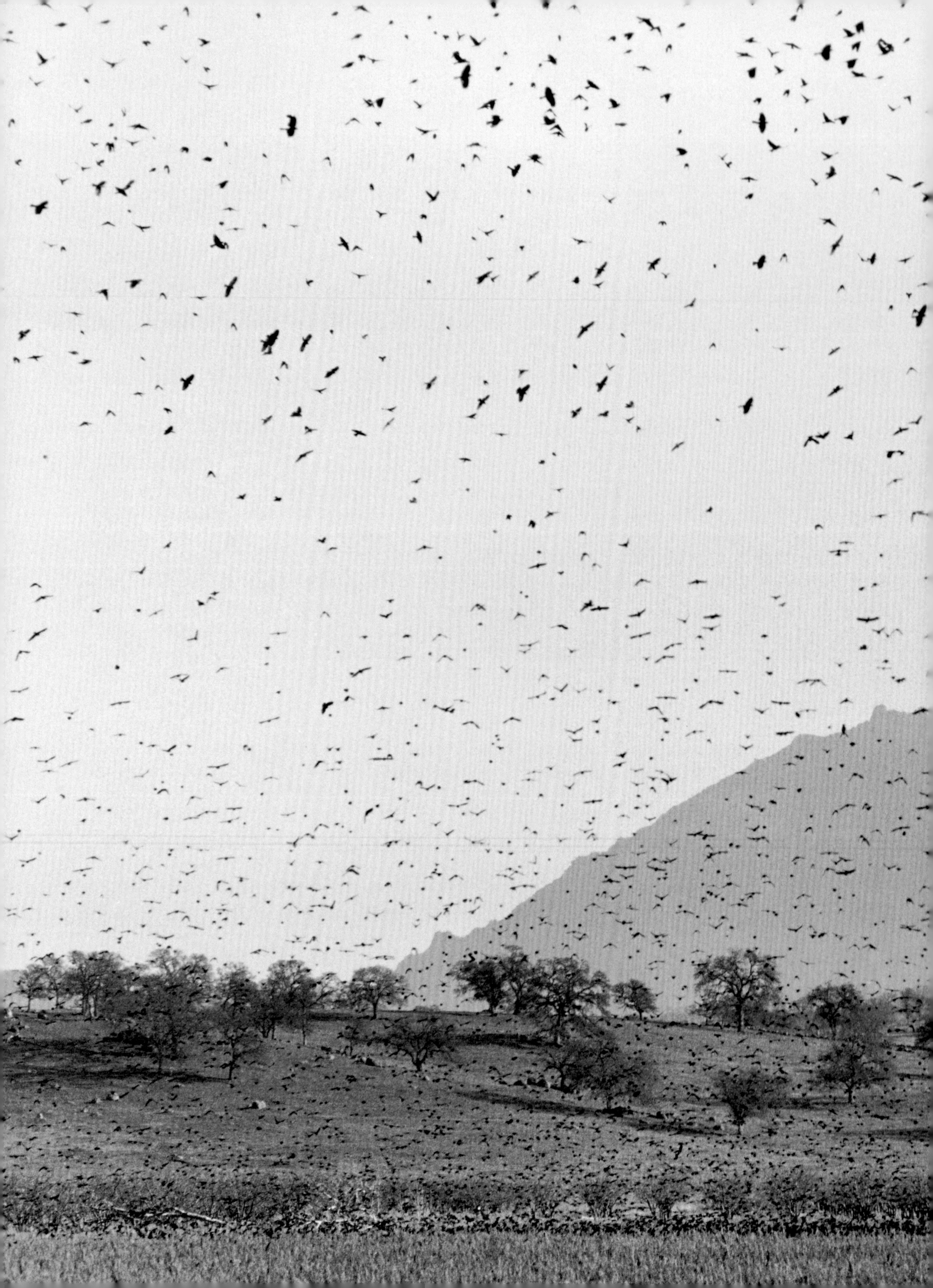

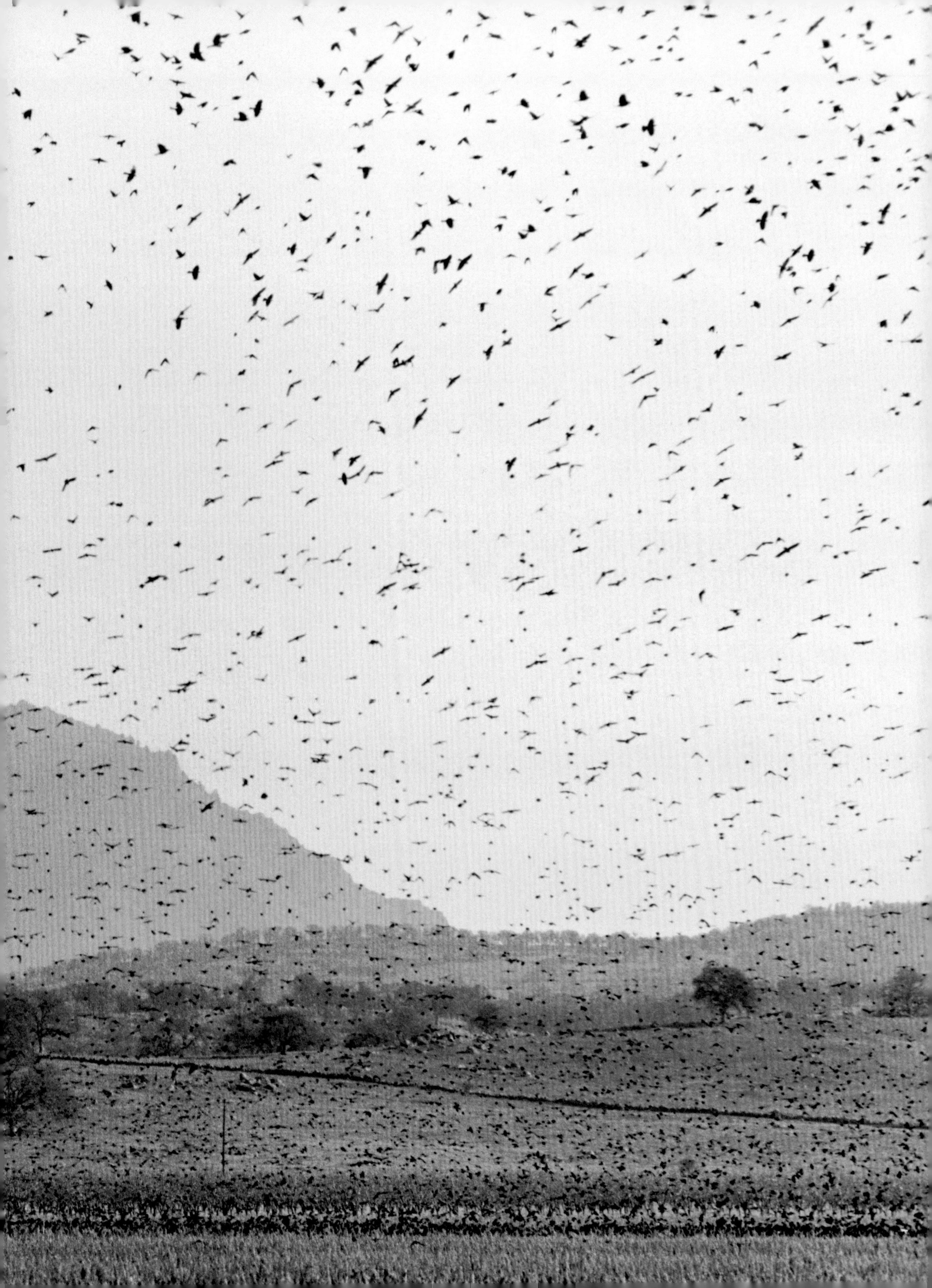

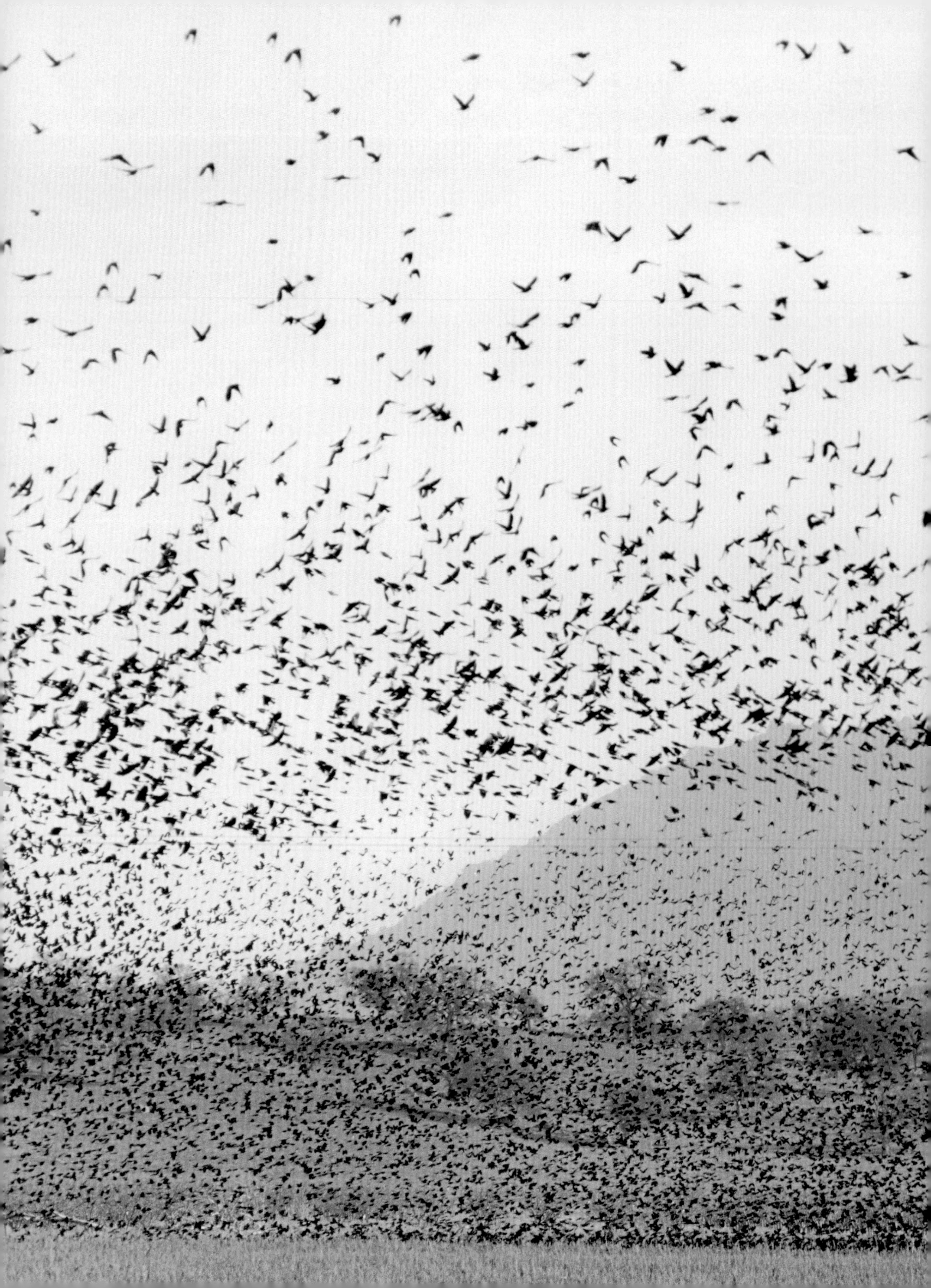

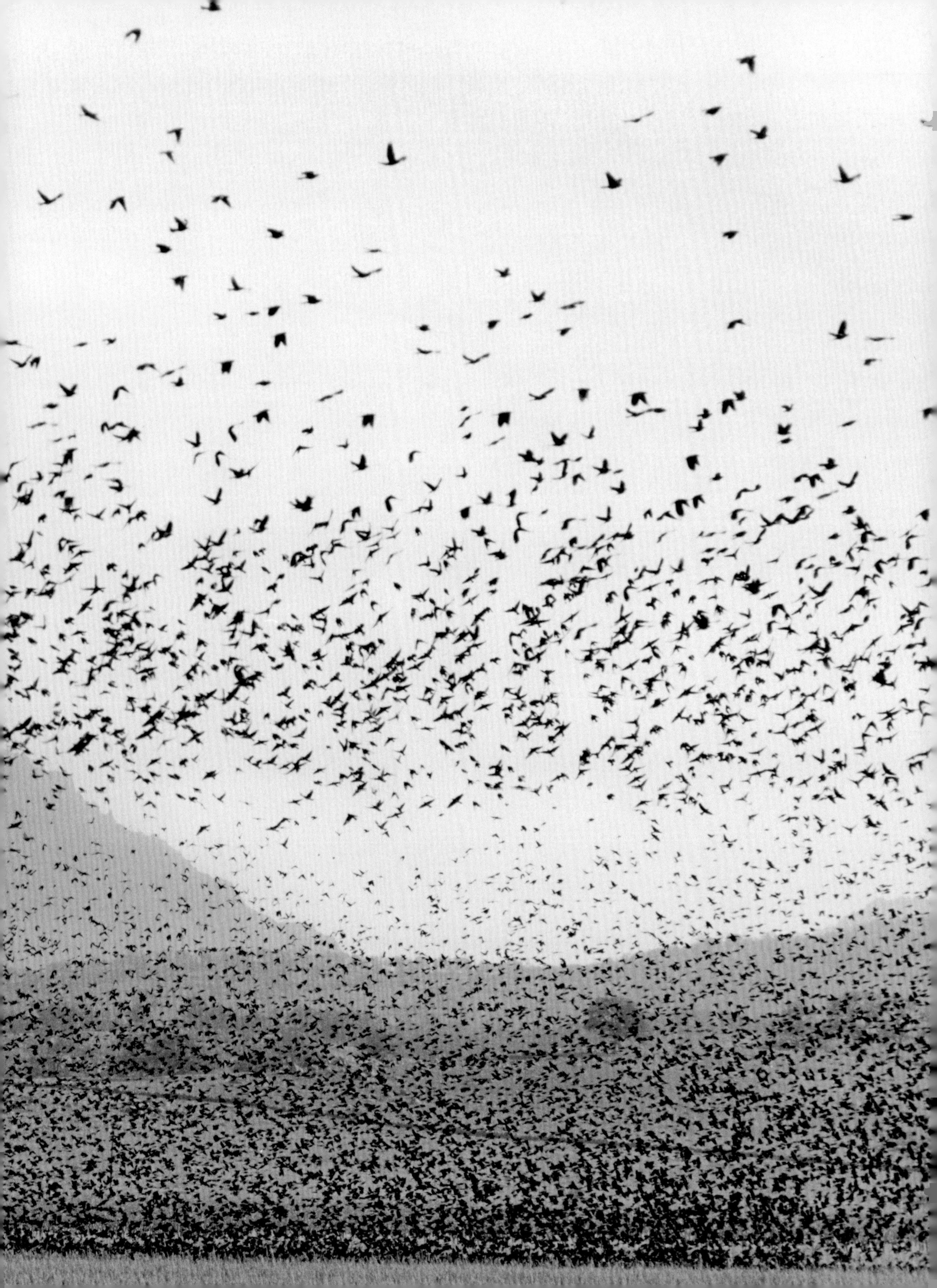

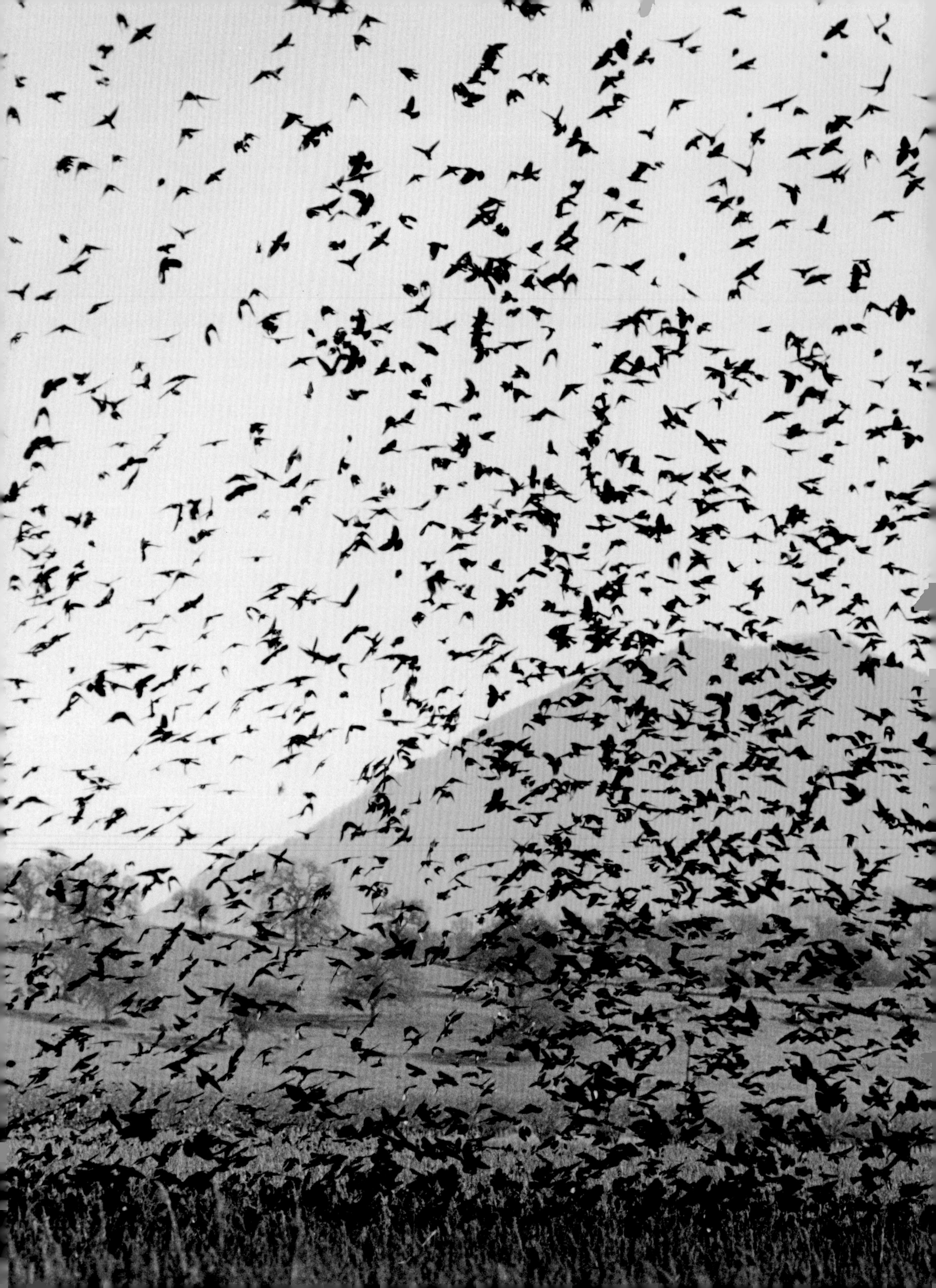

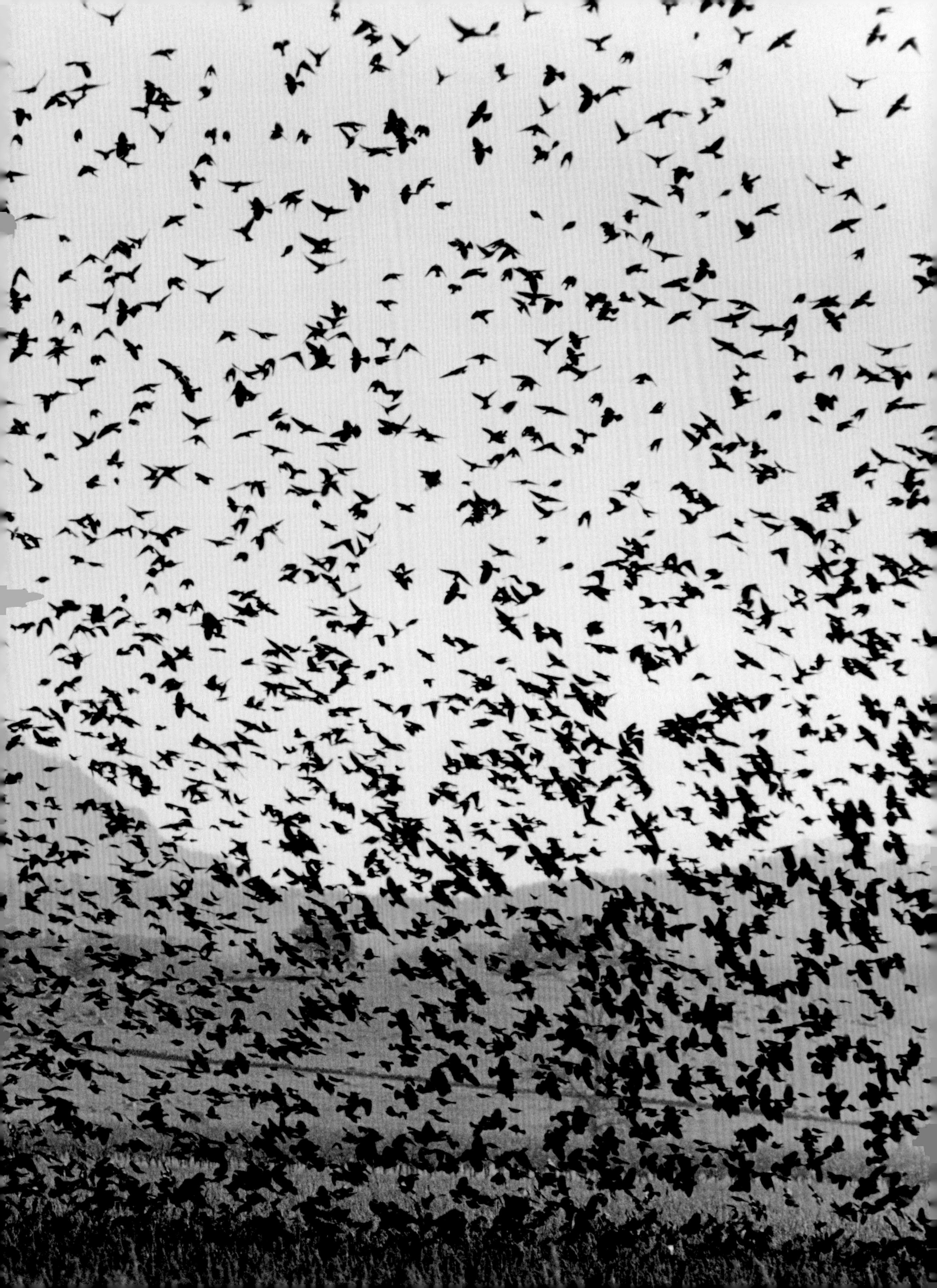

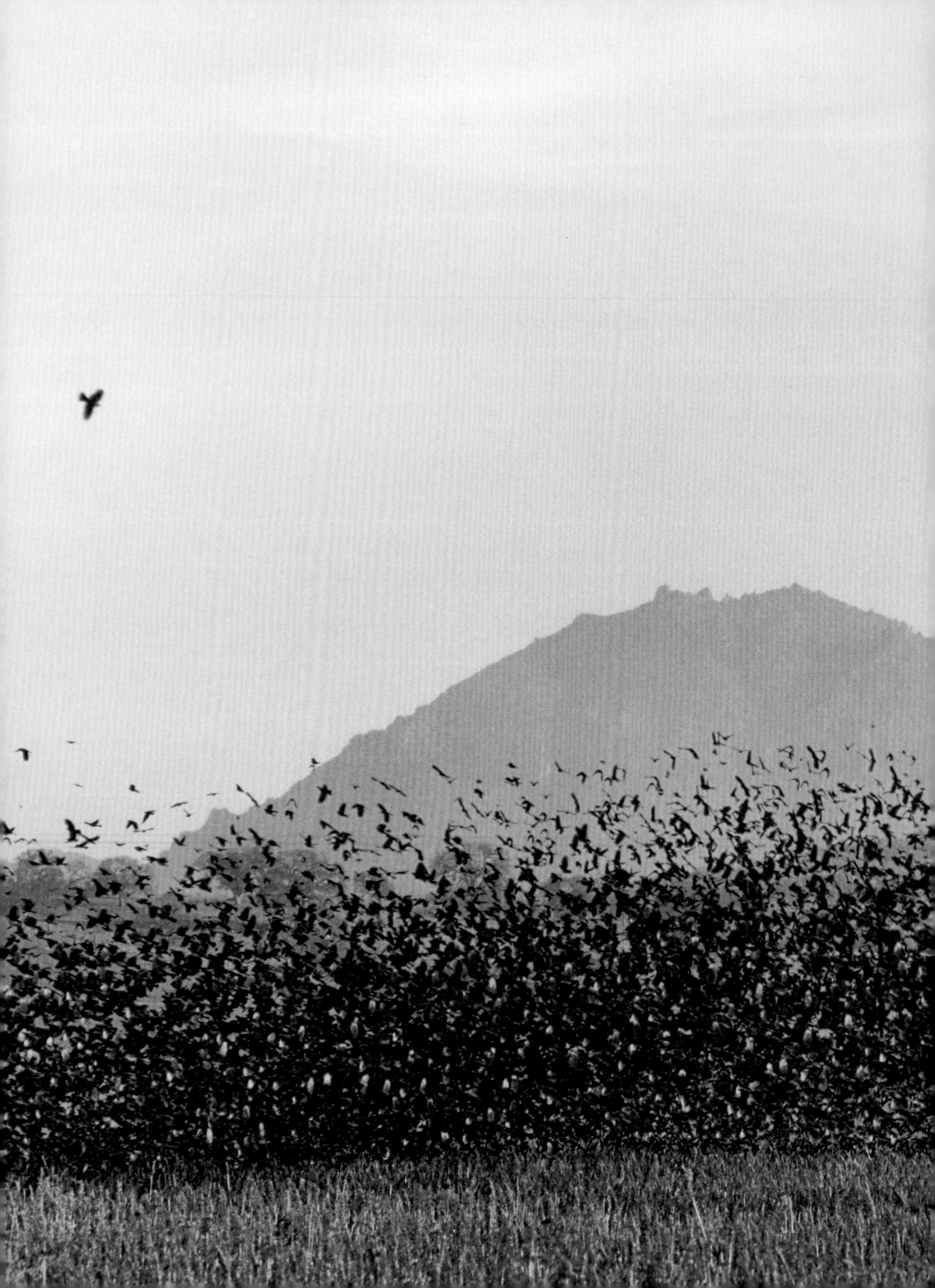

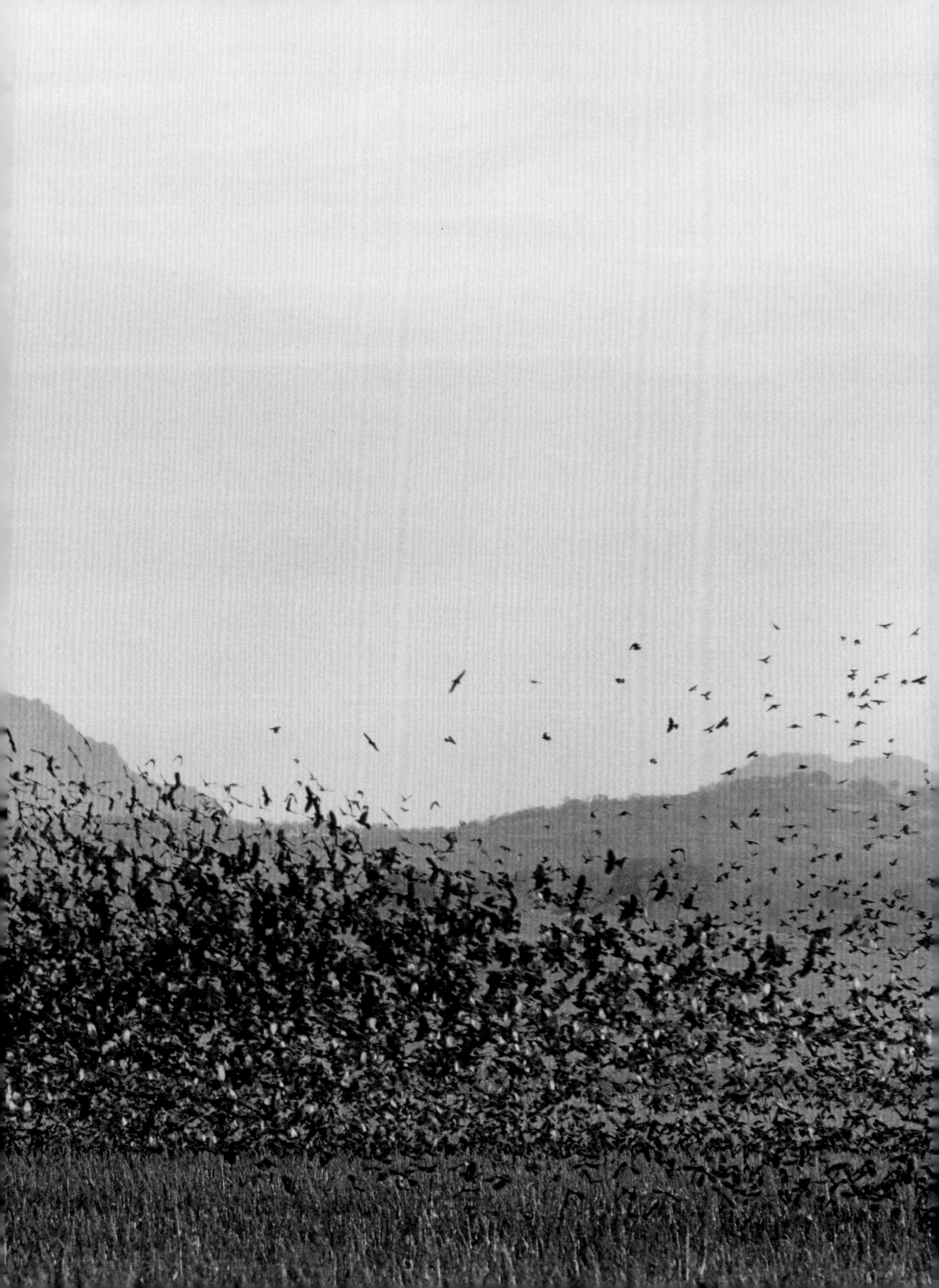

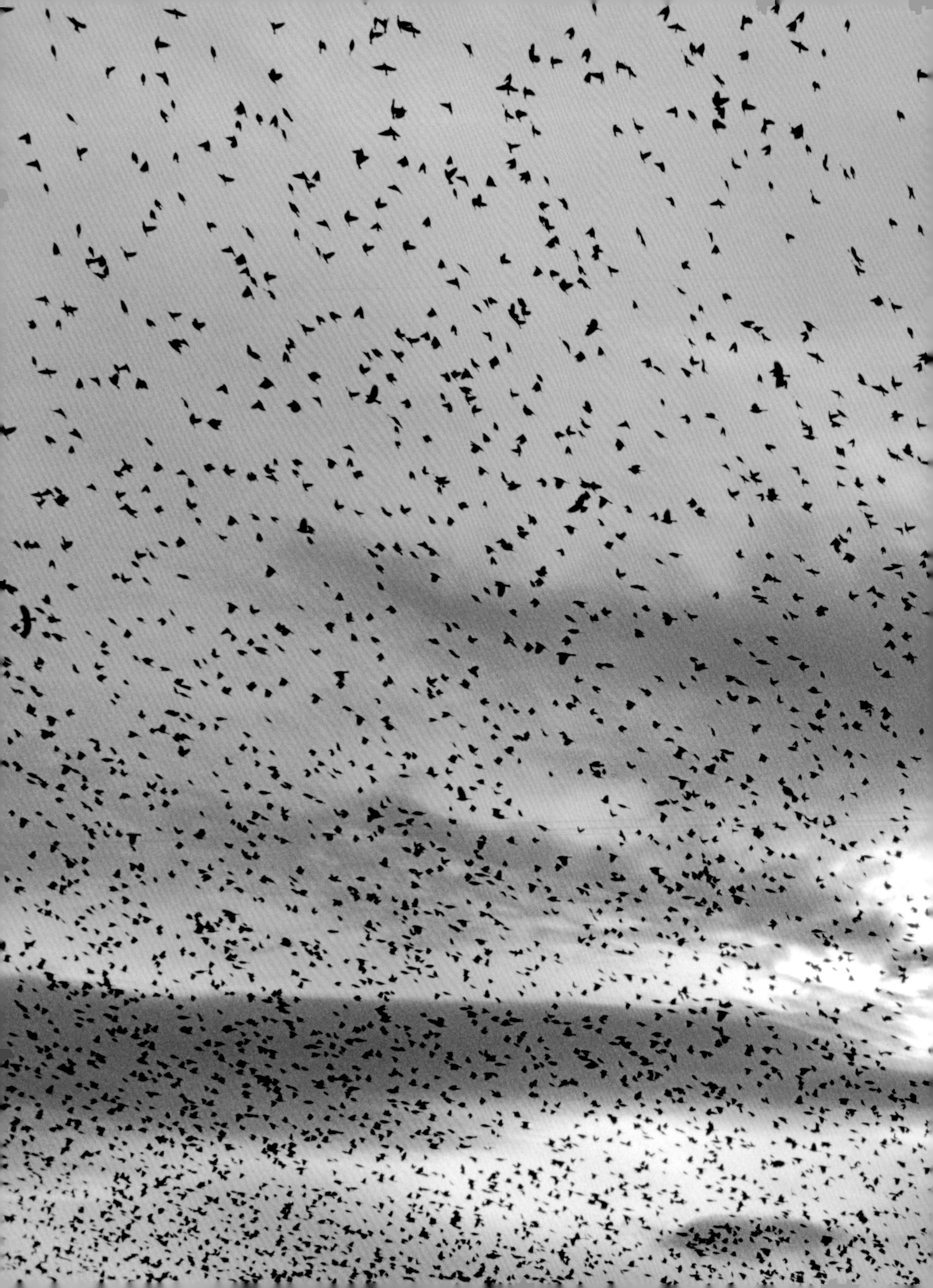

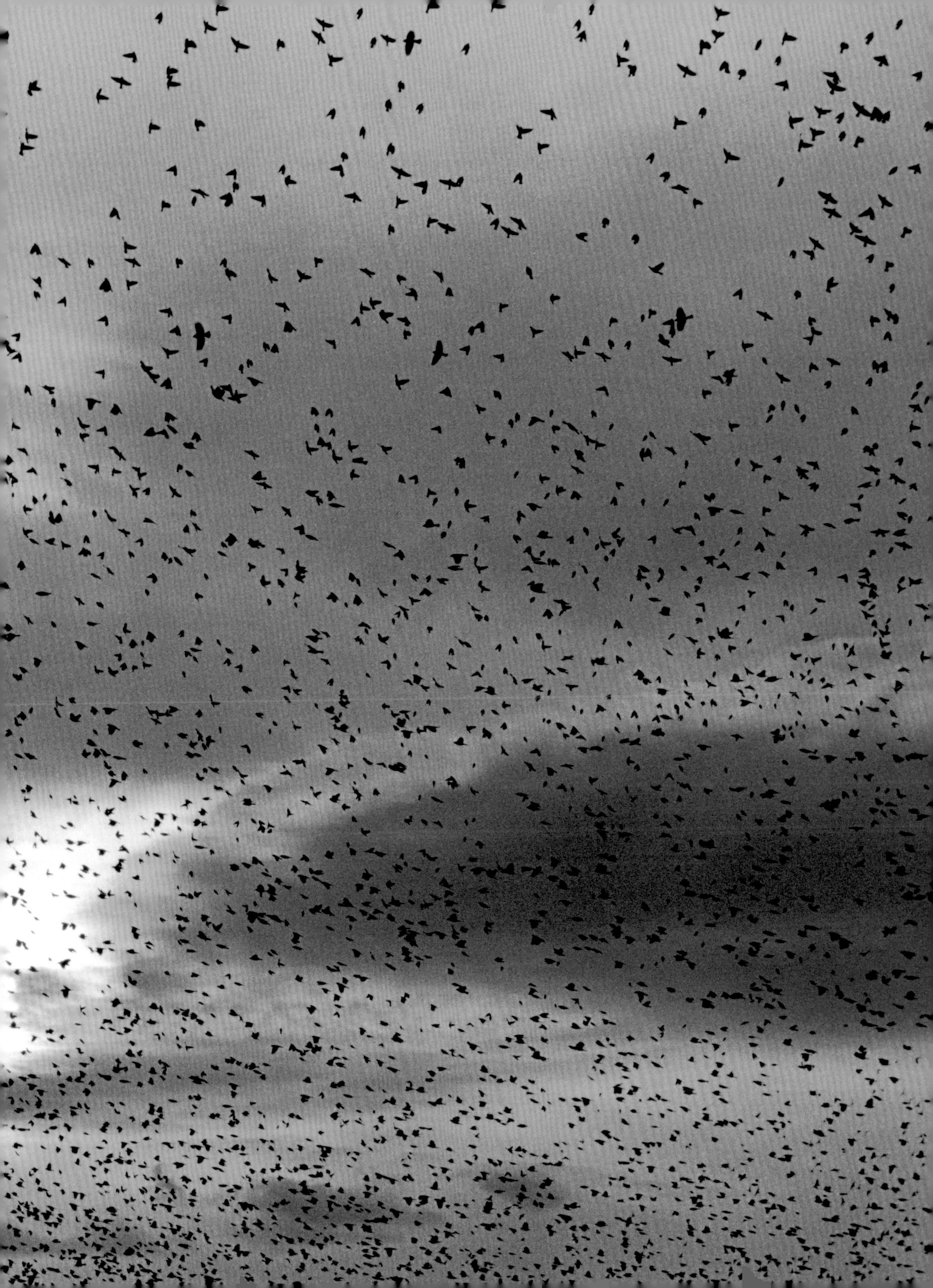

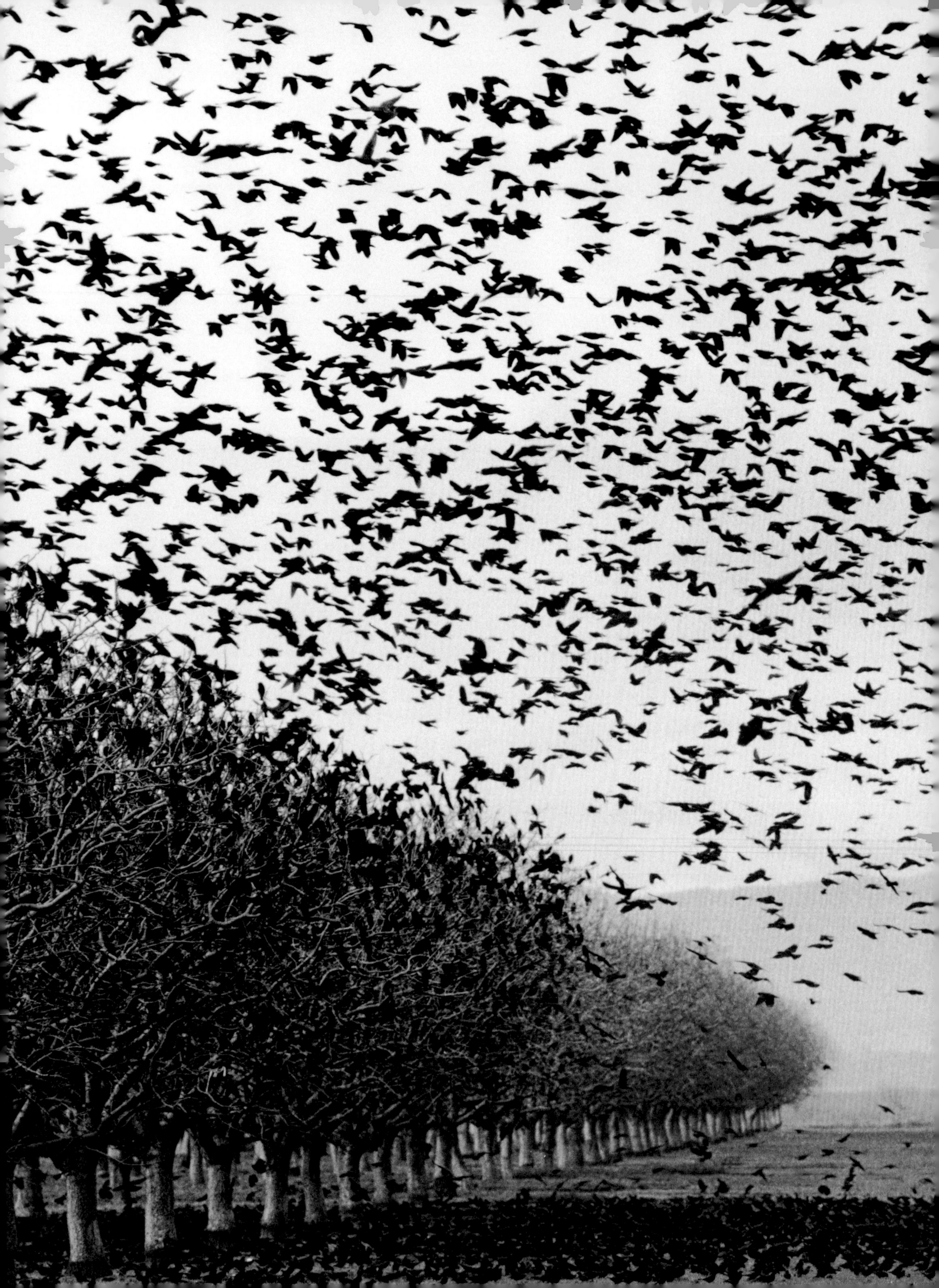

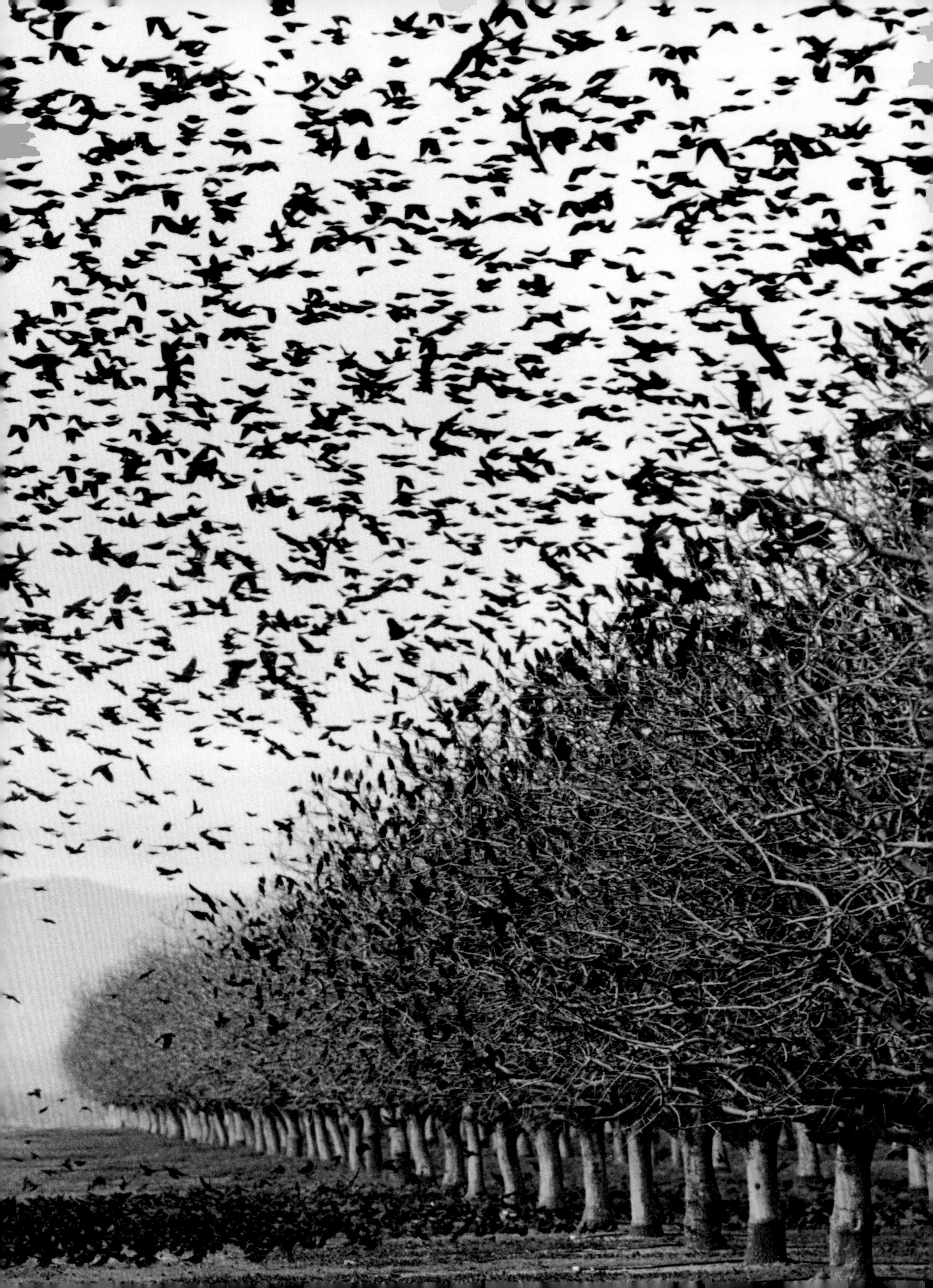

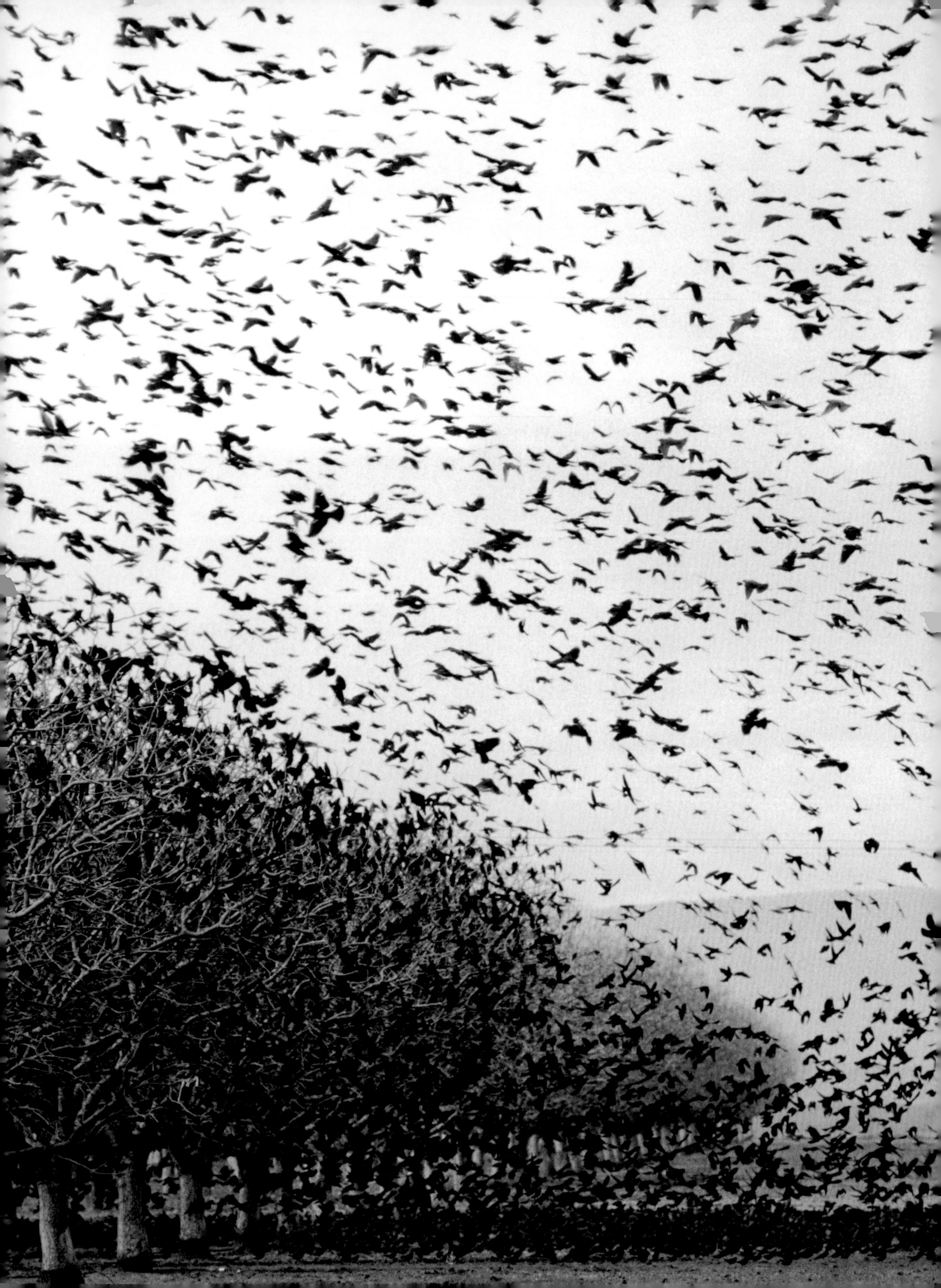

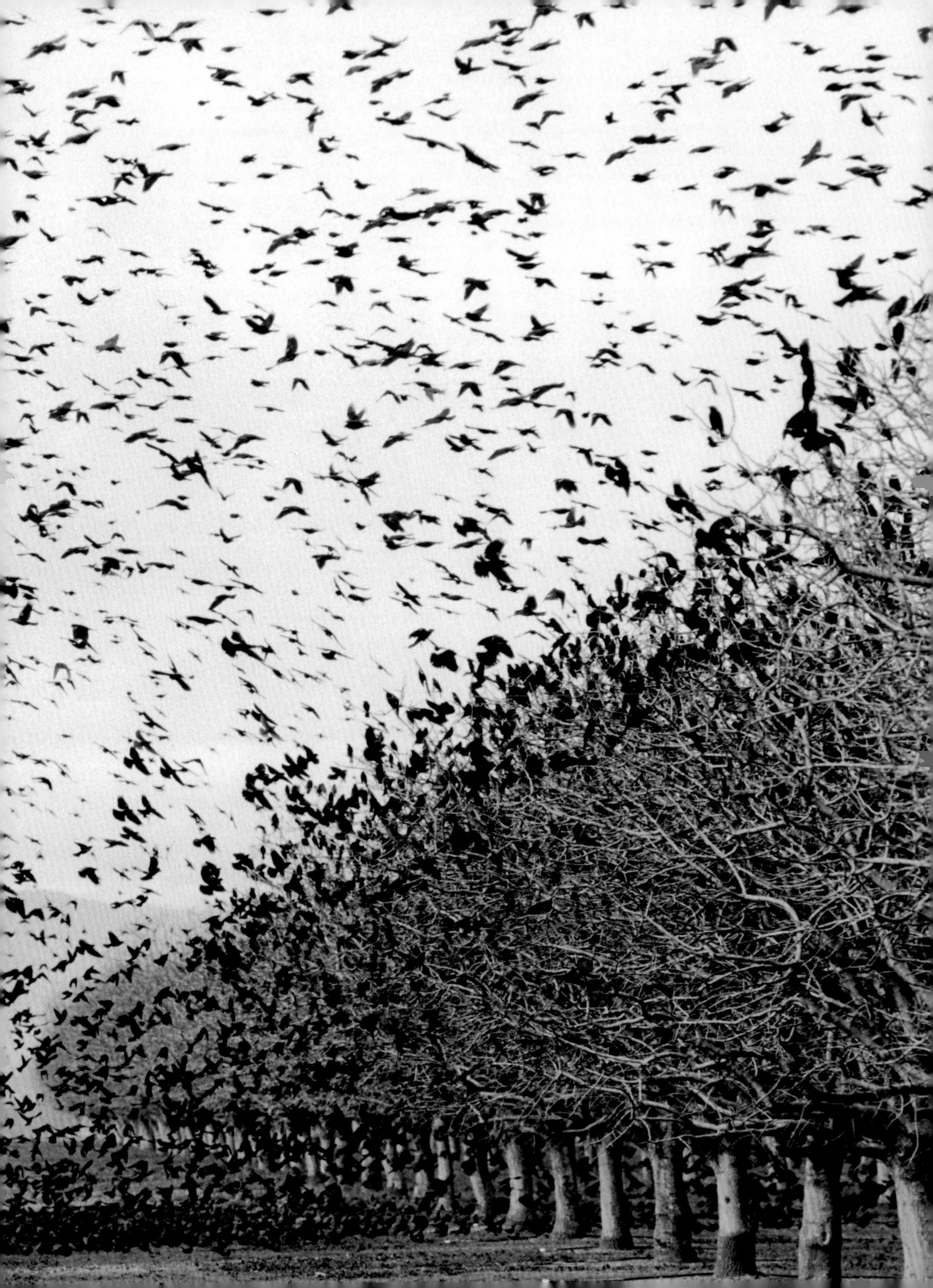

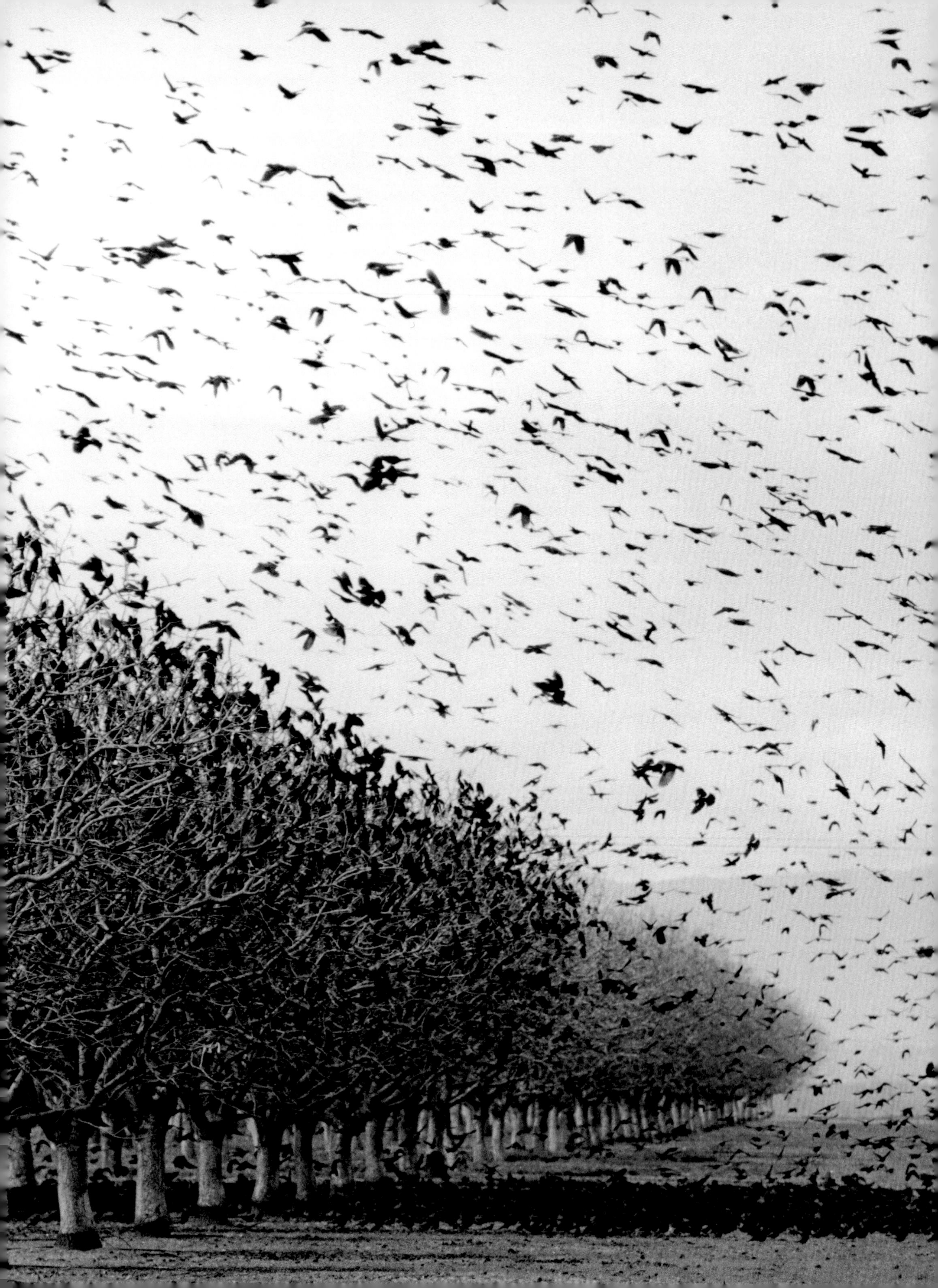

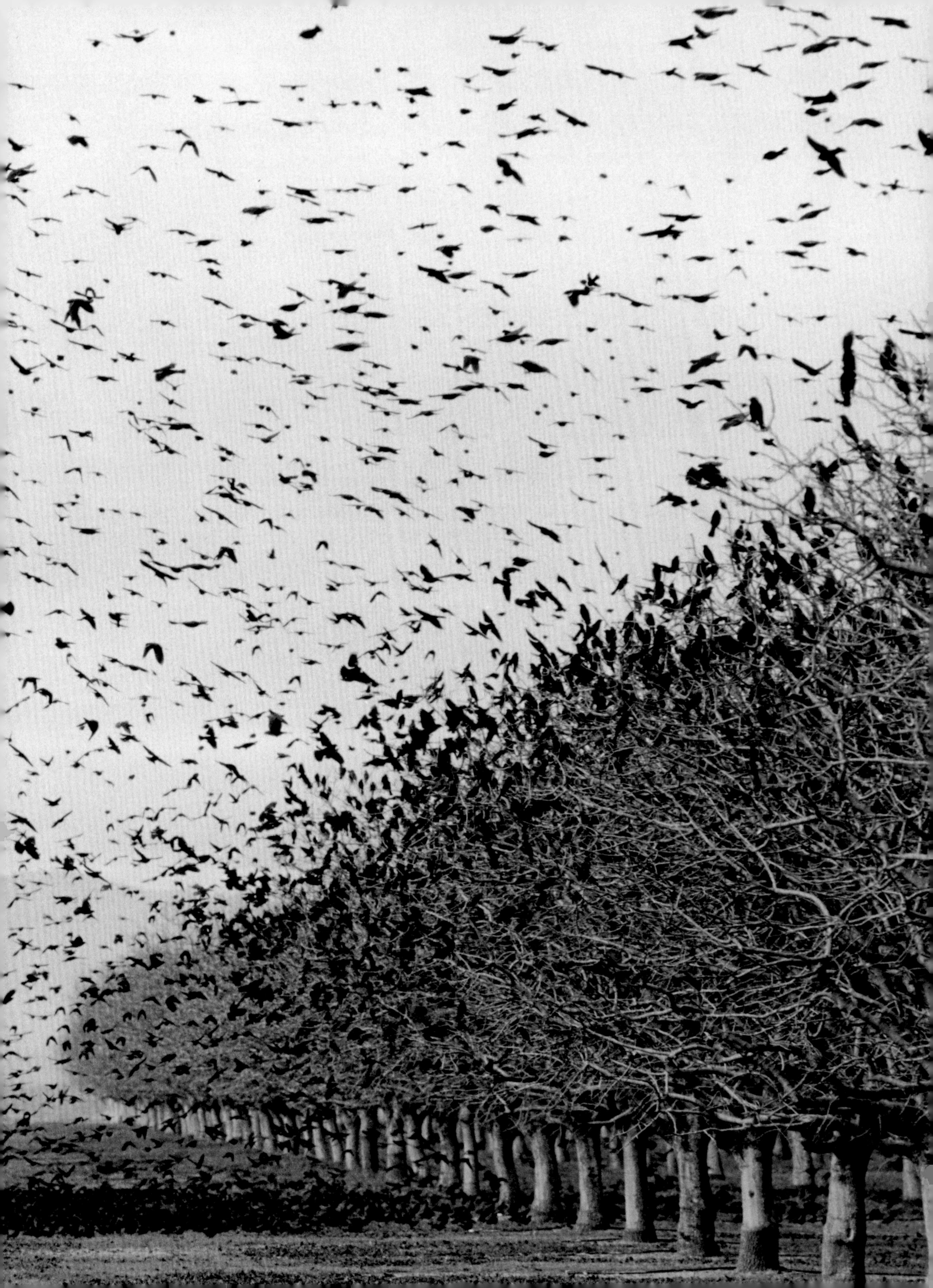

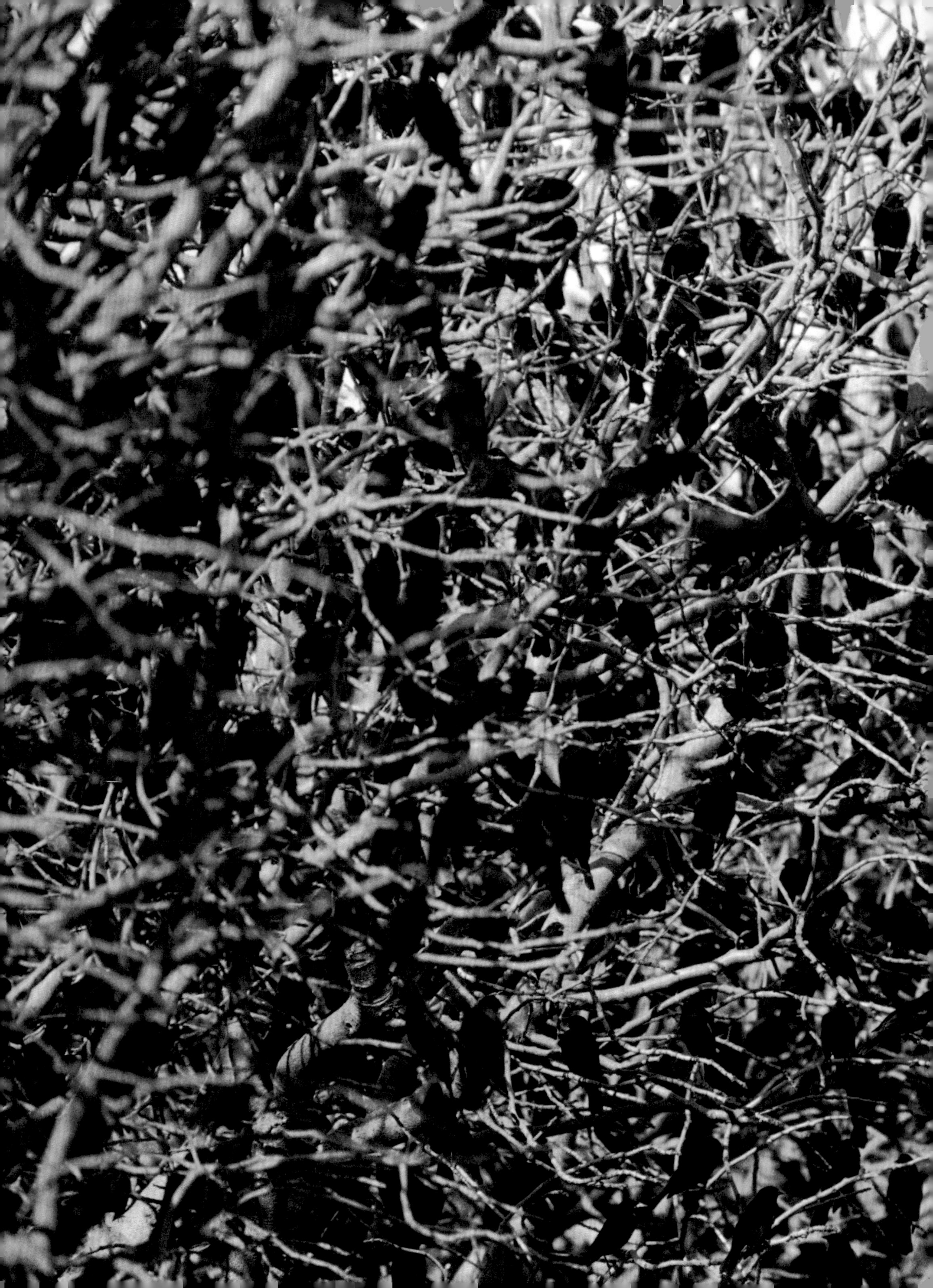

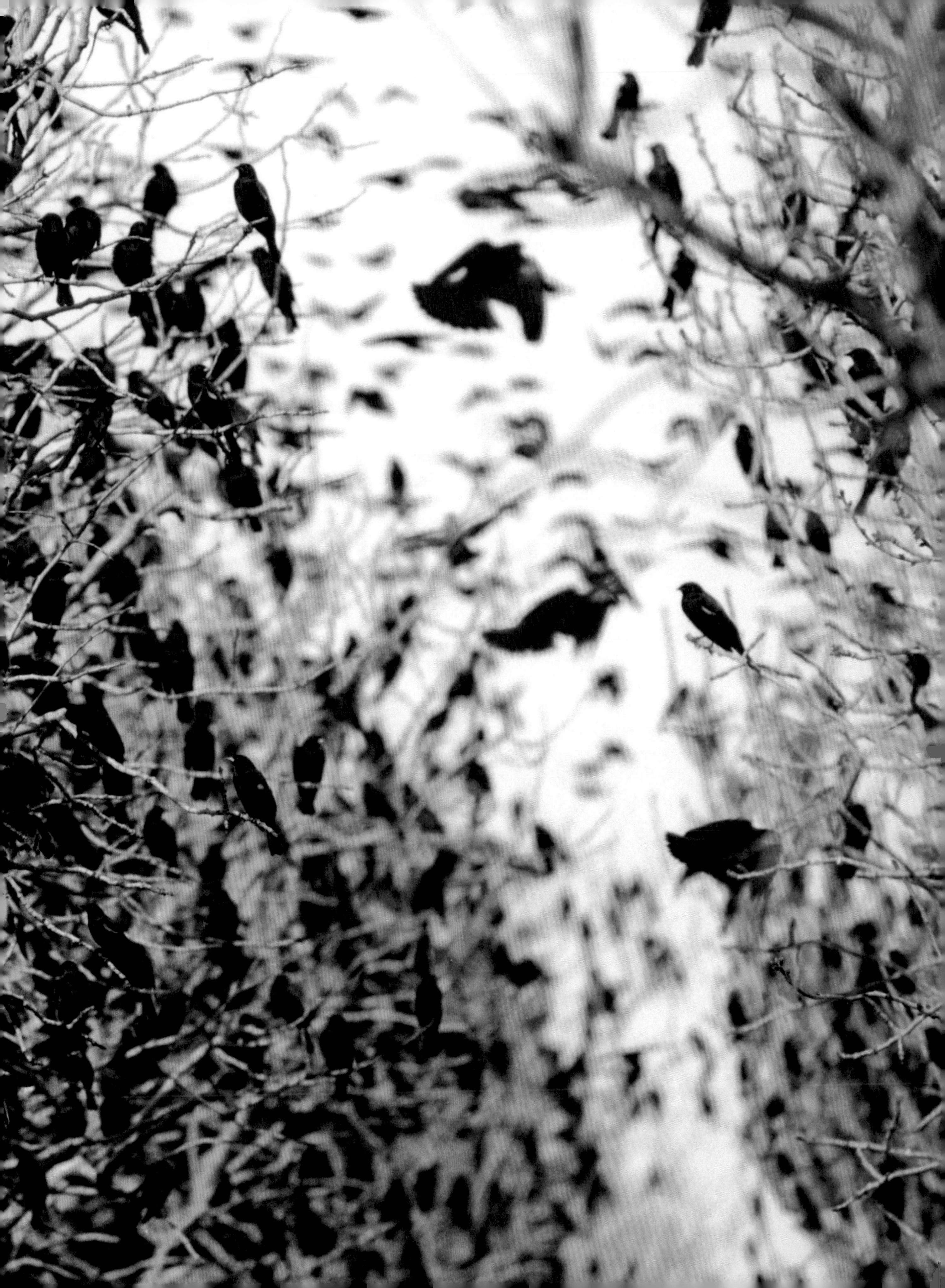

Thirteen Ways of Looking at a Blackbird

Wallace Stevens

I
Among twenty snowy mountains,
The only moving thing
Was the eye of the blackbird.

II
I was of three minds,
Like a tree
In which there are three blackbirds.

III
The blackbird whirled in the autumn winds.
It was a small part of the pantomime.

IV
A man and a woman
Are one.
A man and a woman and a blackbird
Are one.

V
I do not know which to prefer,
The beauty of inflections
Or the beauty of innuendoes,
The blackbird whistling
Or just after.

VI
Icicles filled the long window
With barbaric glass.
The shadow of the blackbird
Crossed it, to and fro.
The mood
Traced in the shadow
An indecipherable cause.

VII

O thin men of Haddam,
Why do you imagine golden birds?
Do you not see how the blackbird
Walks around the feet
Of the women about you?

VIII

I know noble accents
And lucid, inescapable rhythms;
But I know, too,
That the blackbird is involved
In what I know.

IX

When the blackbird flew out of sight,
It marked the edge
Of one of many circles.

X

At the sight of blackbirds
Flying in a green light,
Even the bawds of euphony
Would cry out sharply.

XI

He rode over Connecticut
In a glass coach.
Once, a fear pierced him,
In that he mistook
The shadow of his equipage
For blackbirds.

XII

The river is moving.
The blackbird must be flying.

XIII

It was evening all afternoon.
It was snowing
And it was going to snow.
The blackbird sat
In the cedar-limbs.

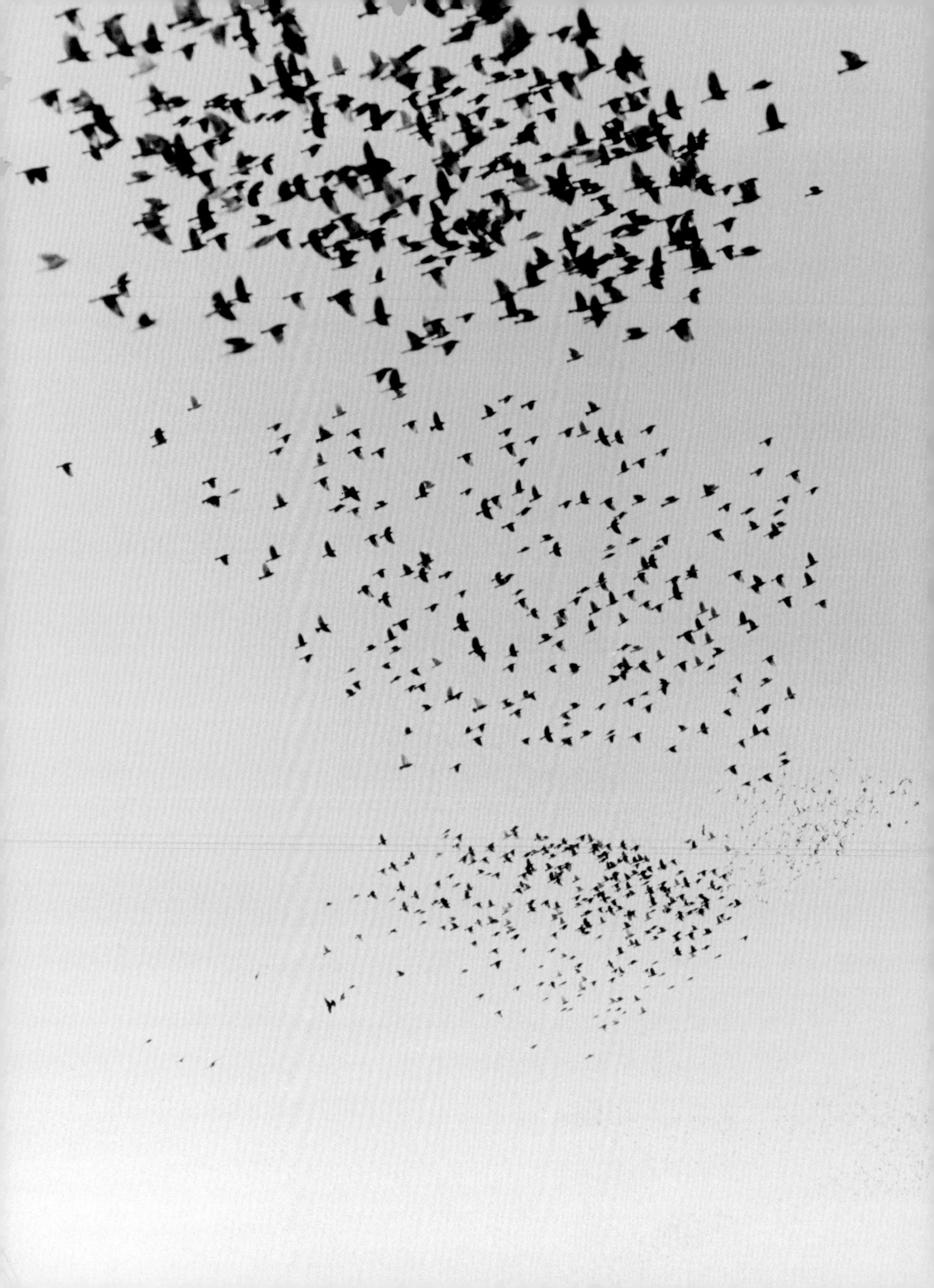

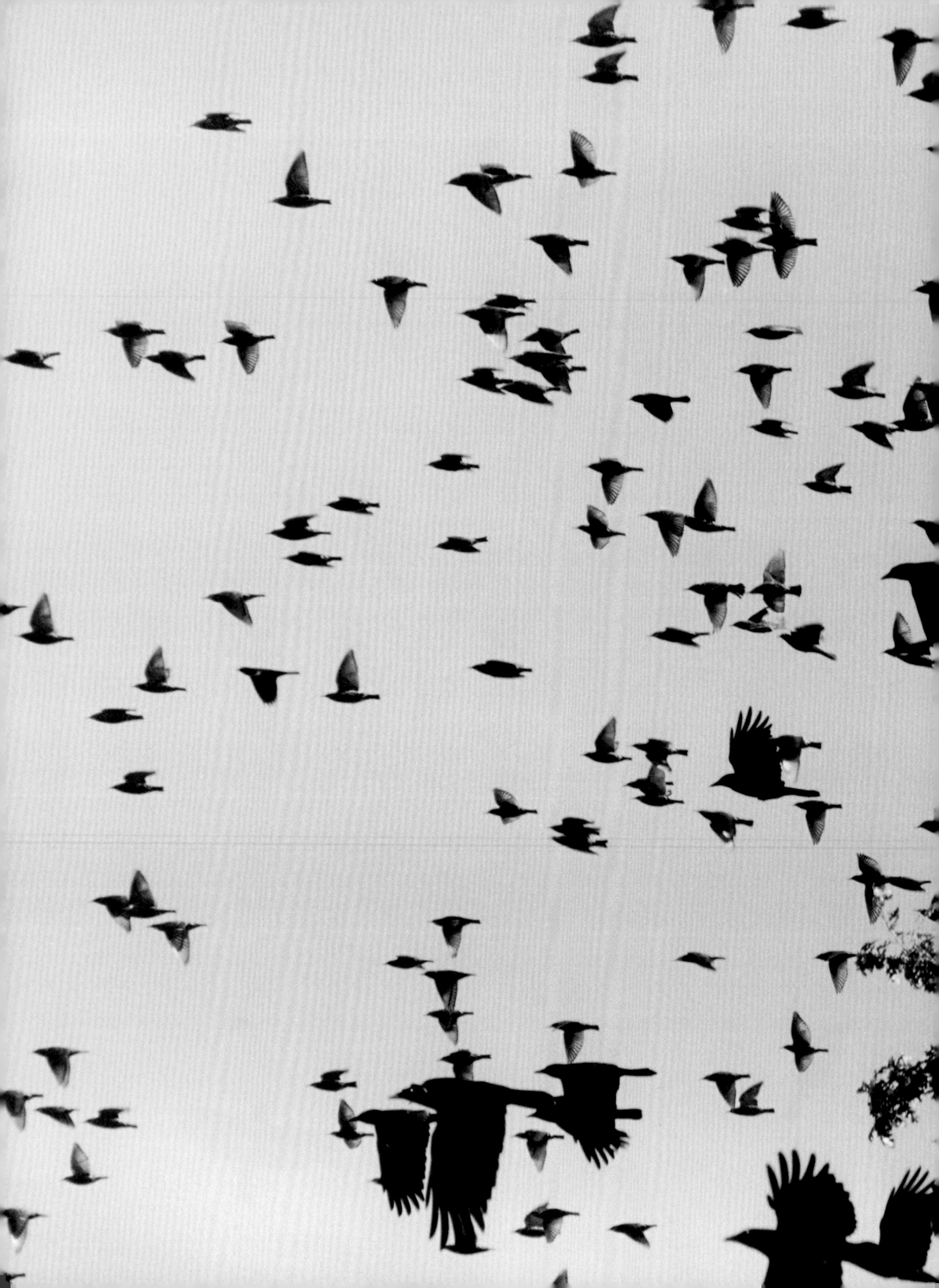

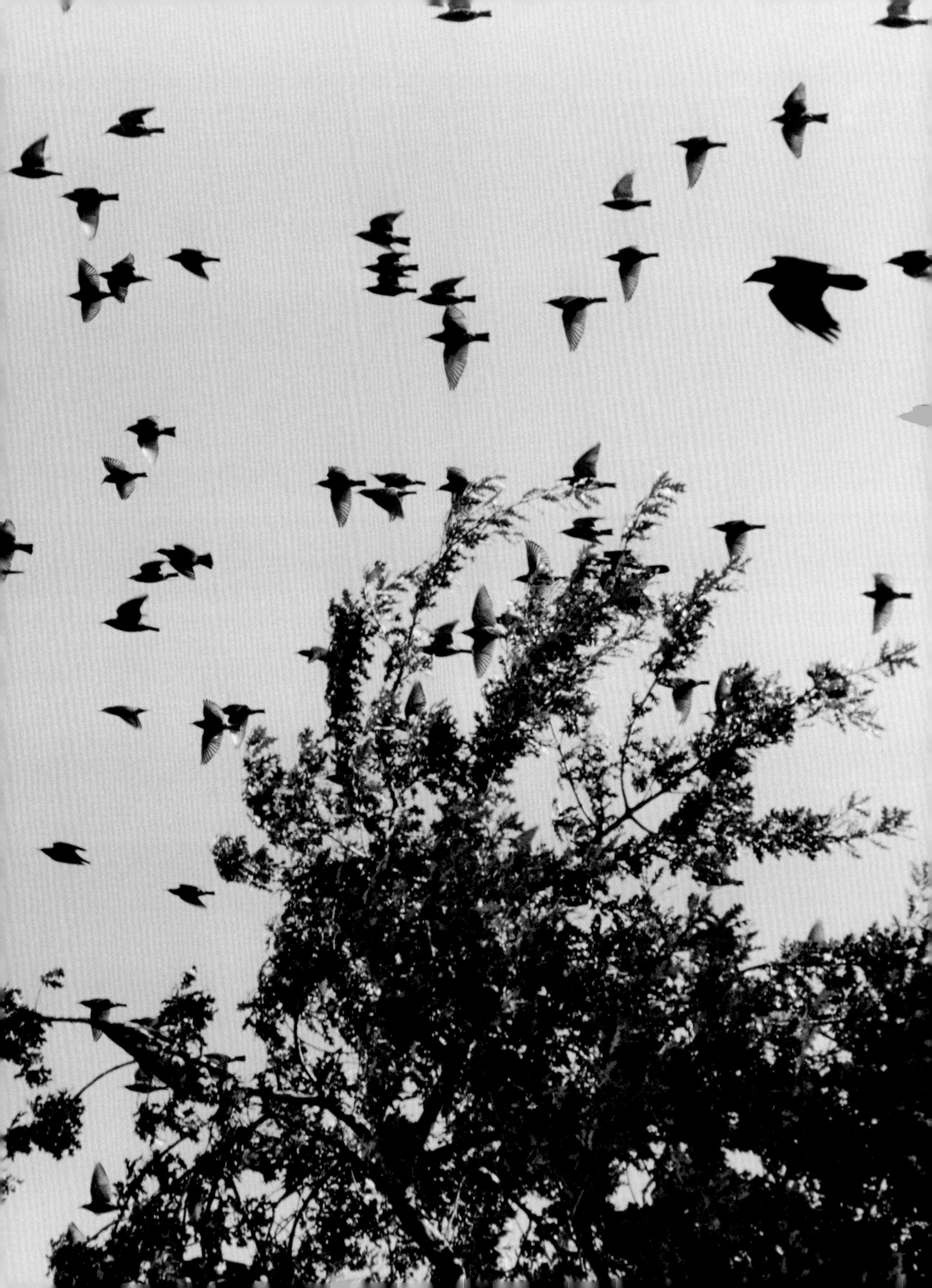

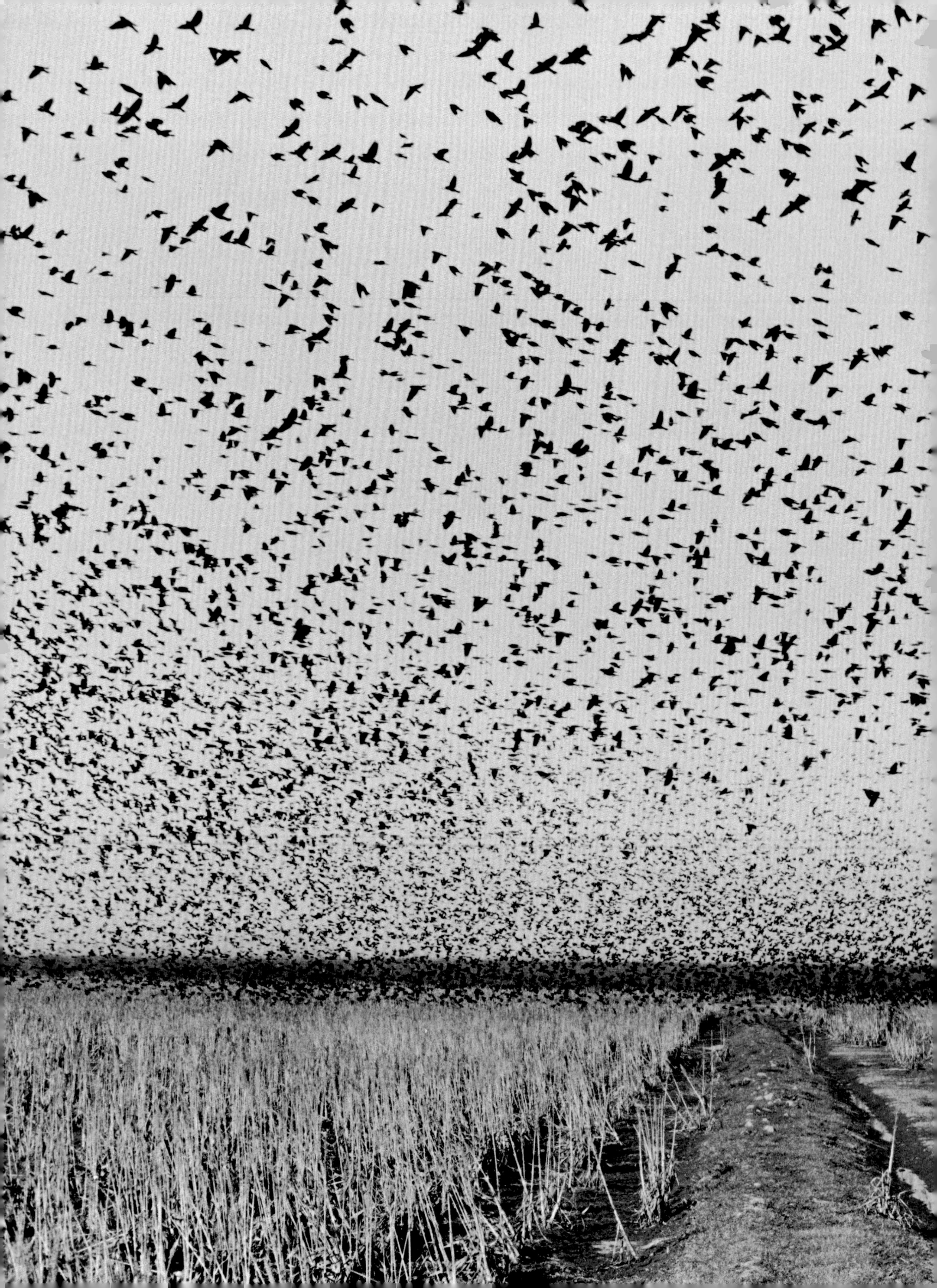

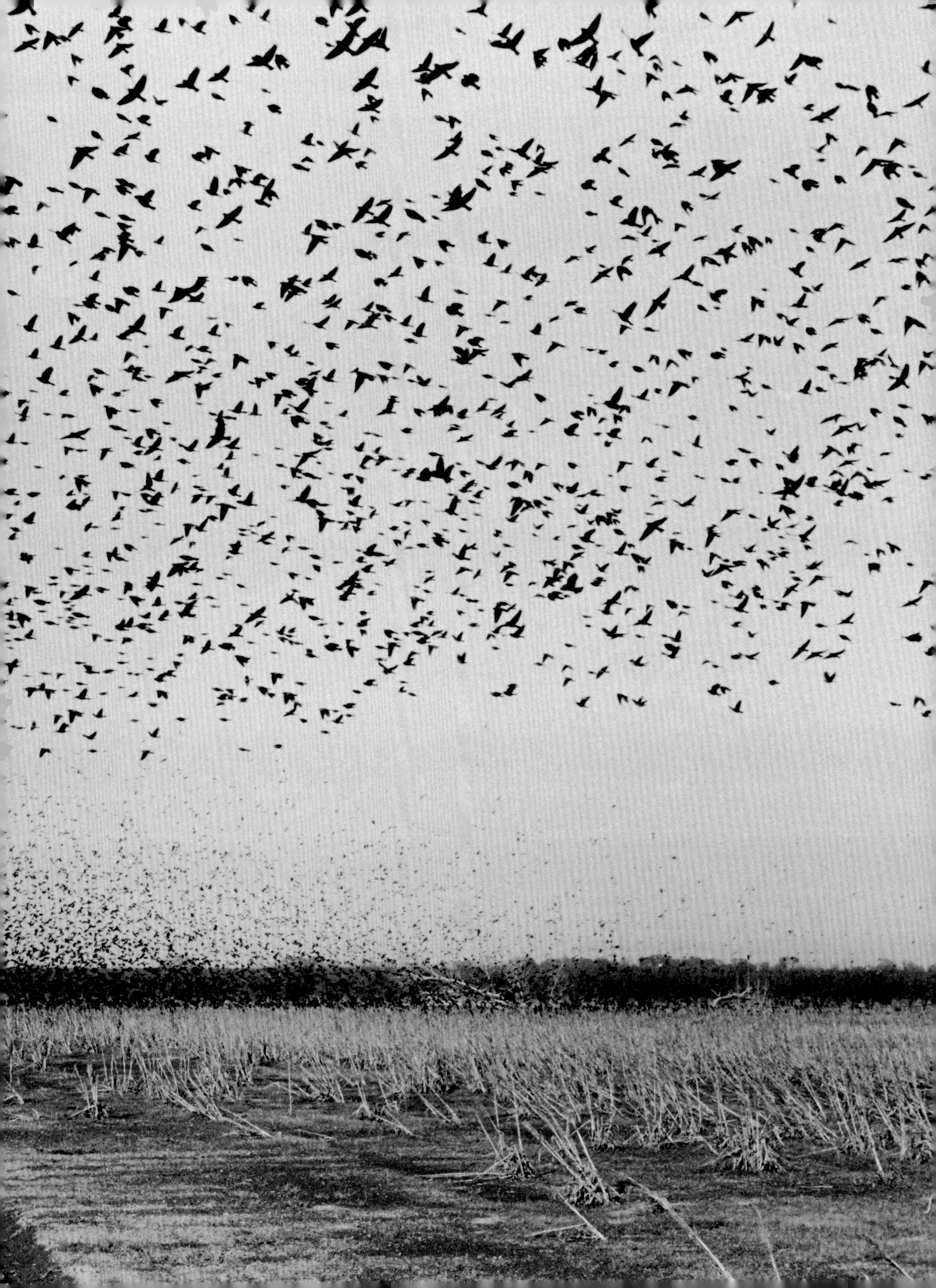

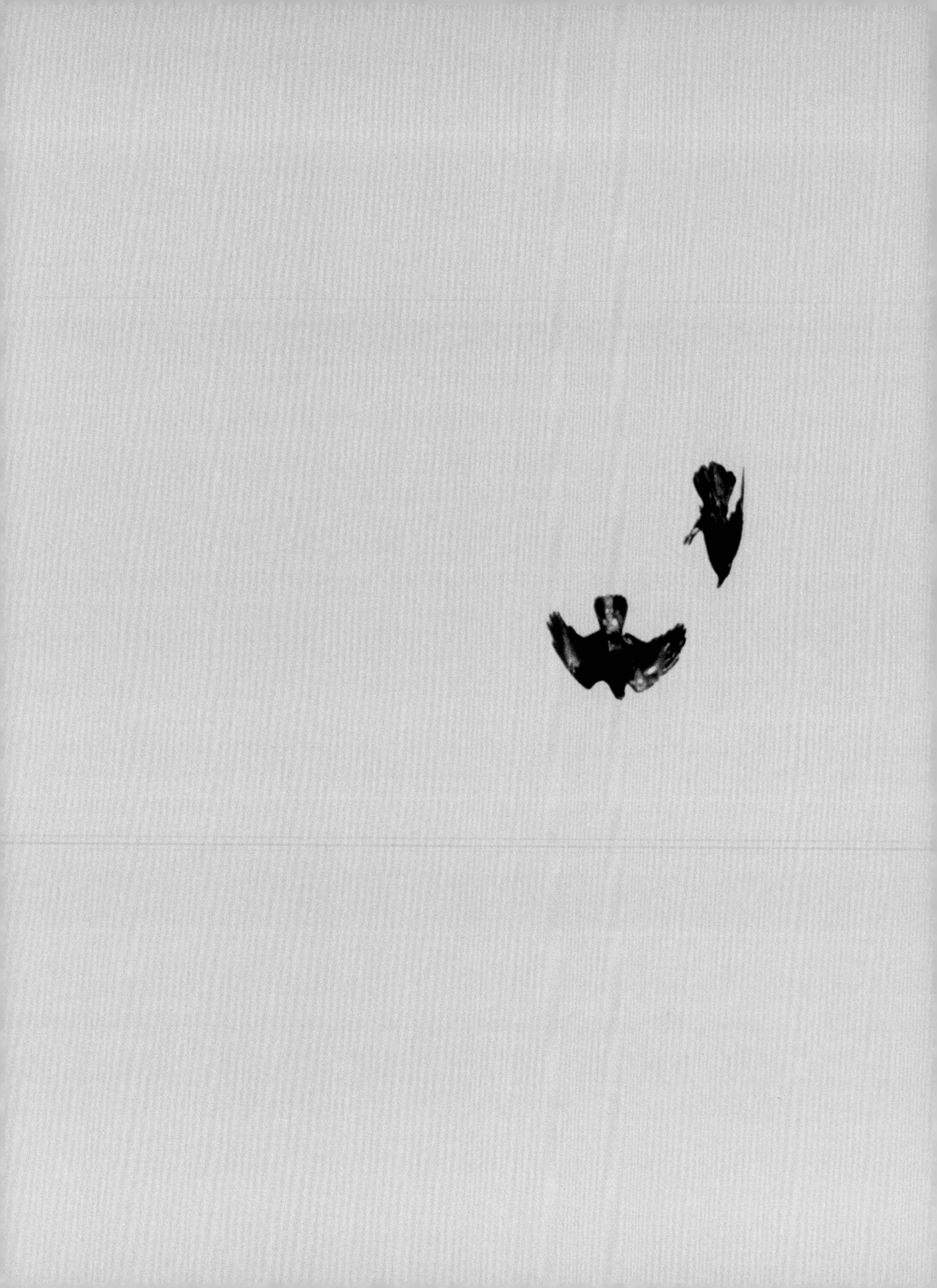

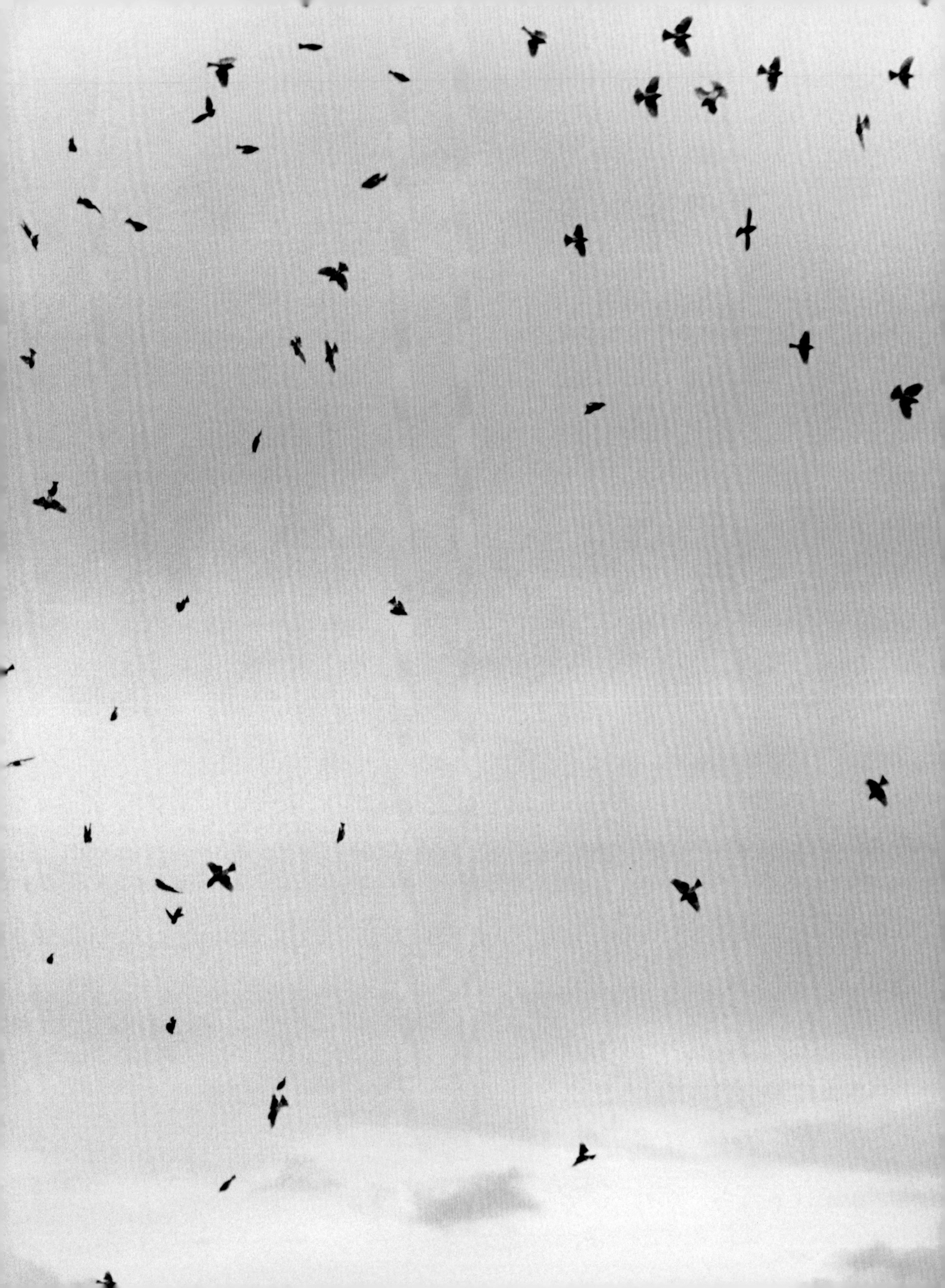

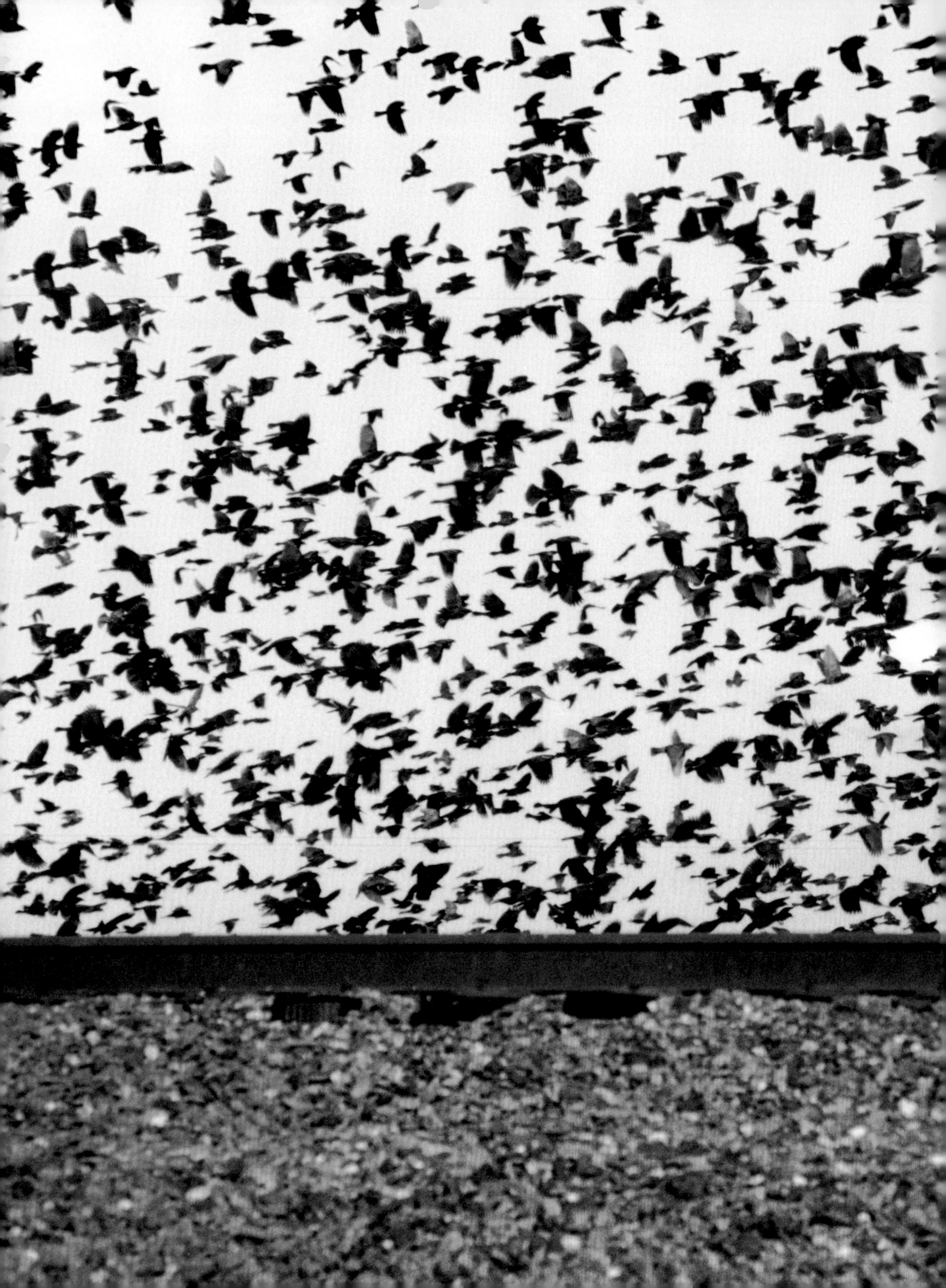

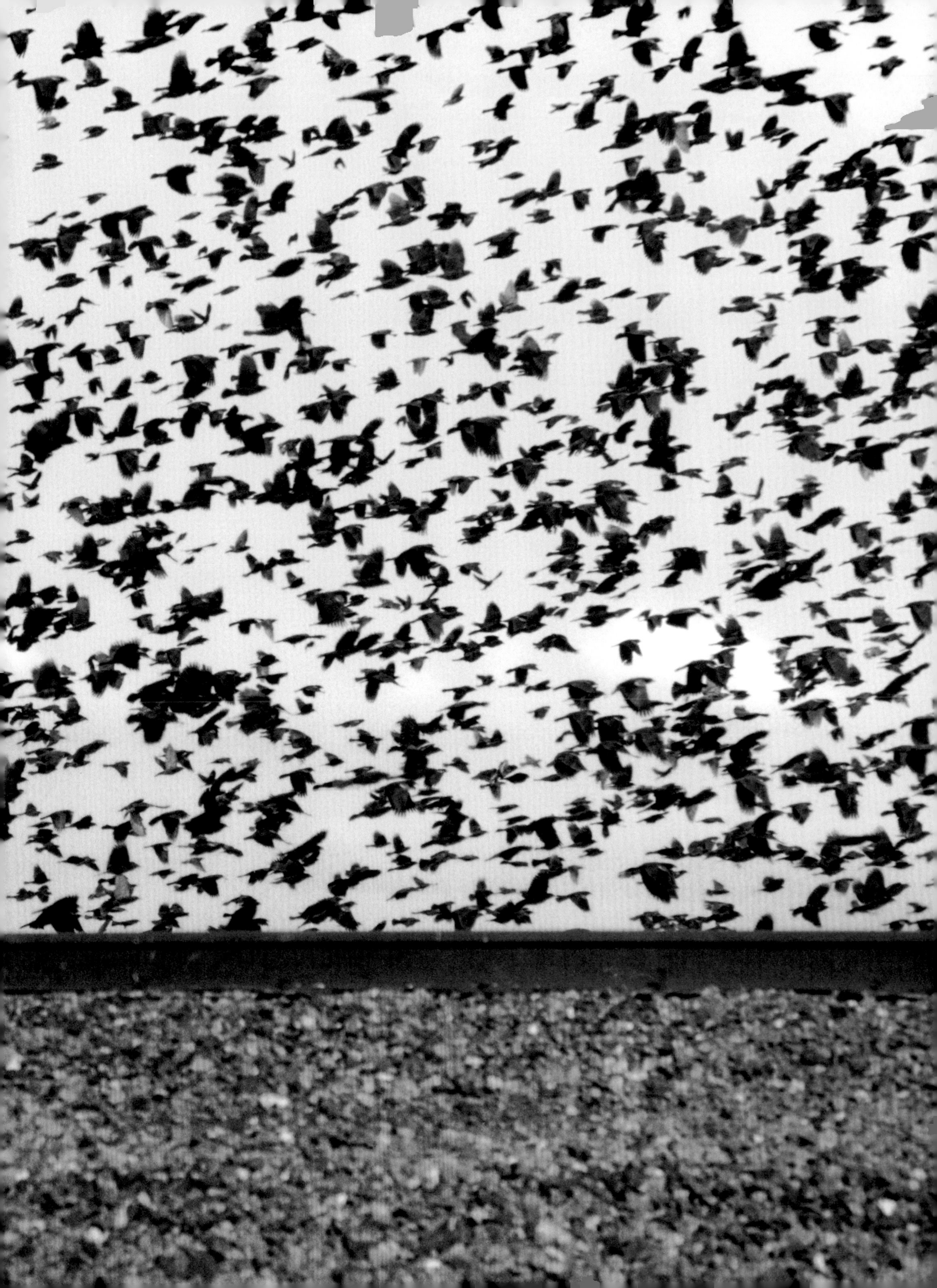

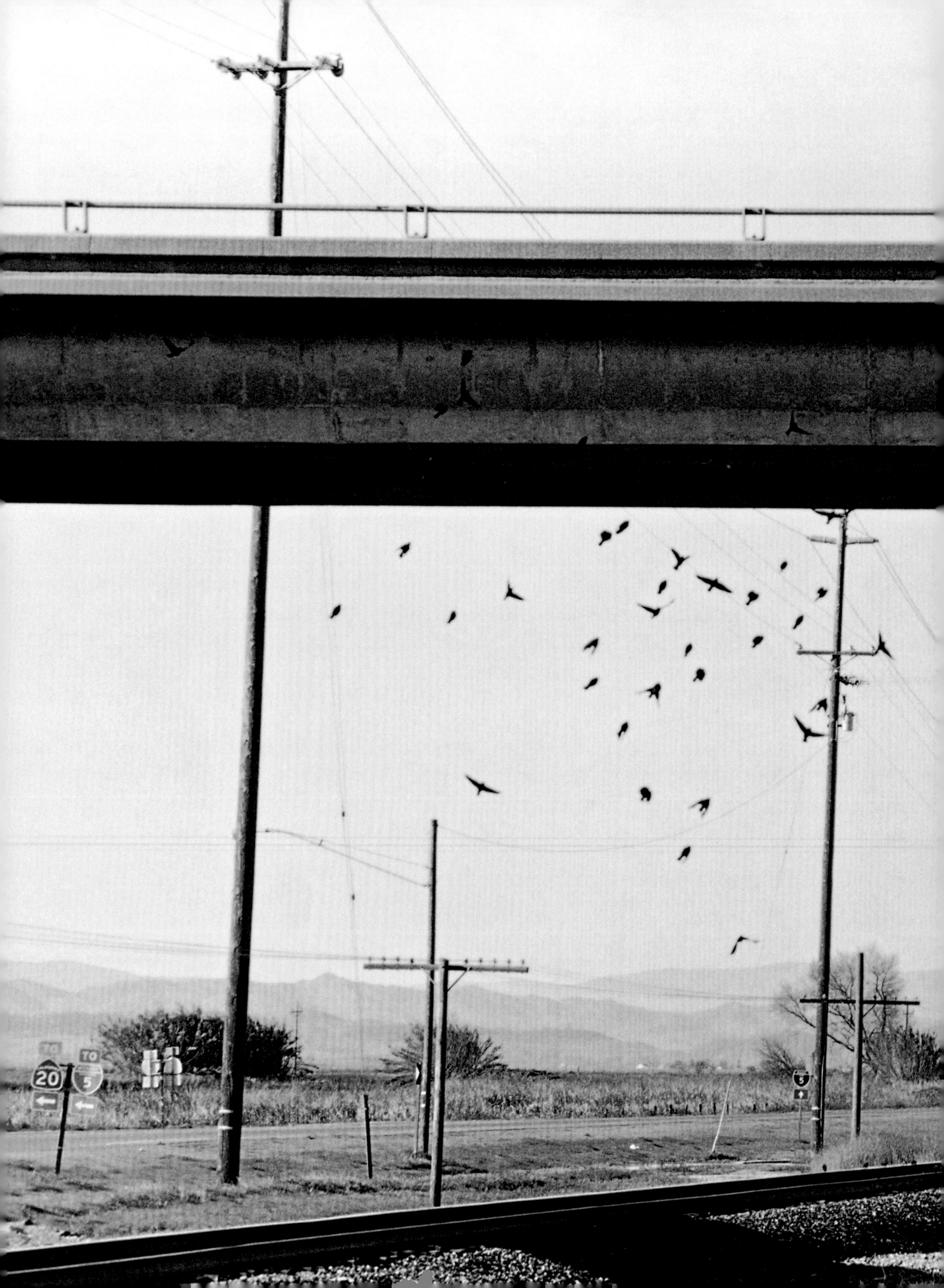
TO
20
TO
5
5

50

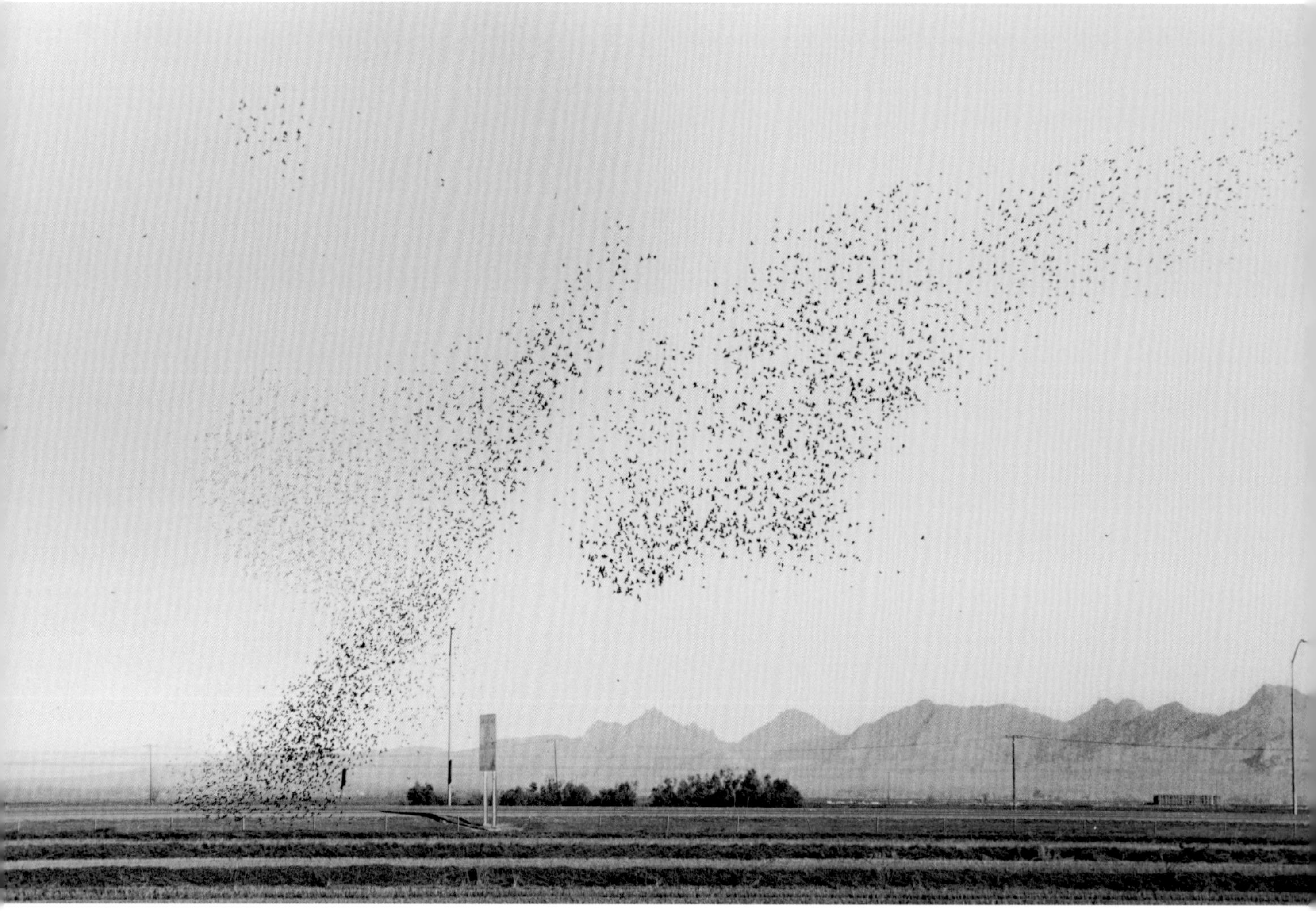

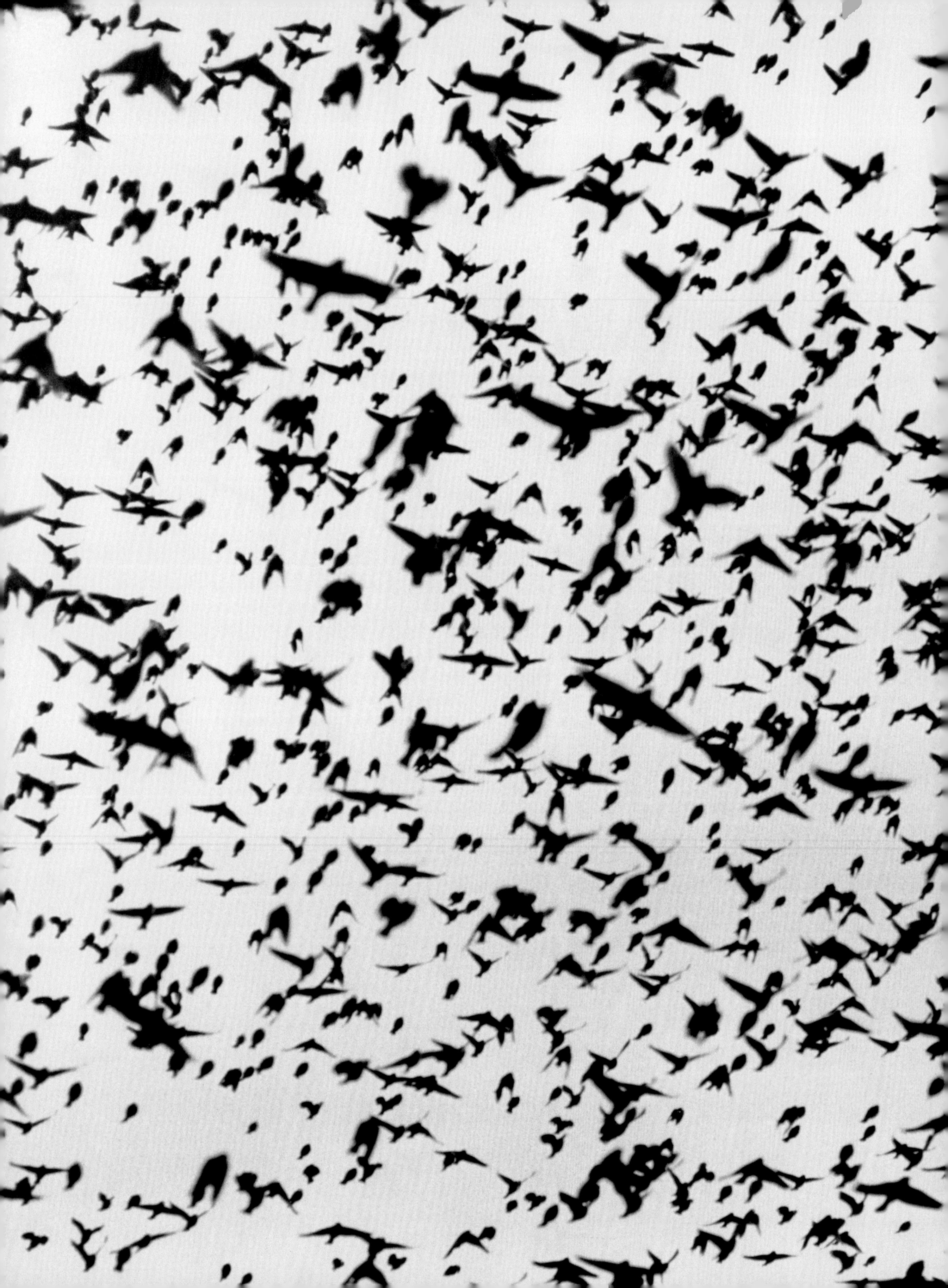

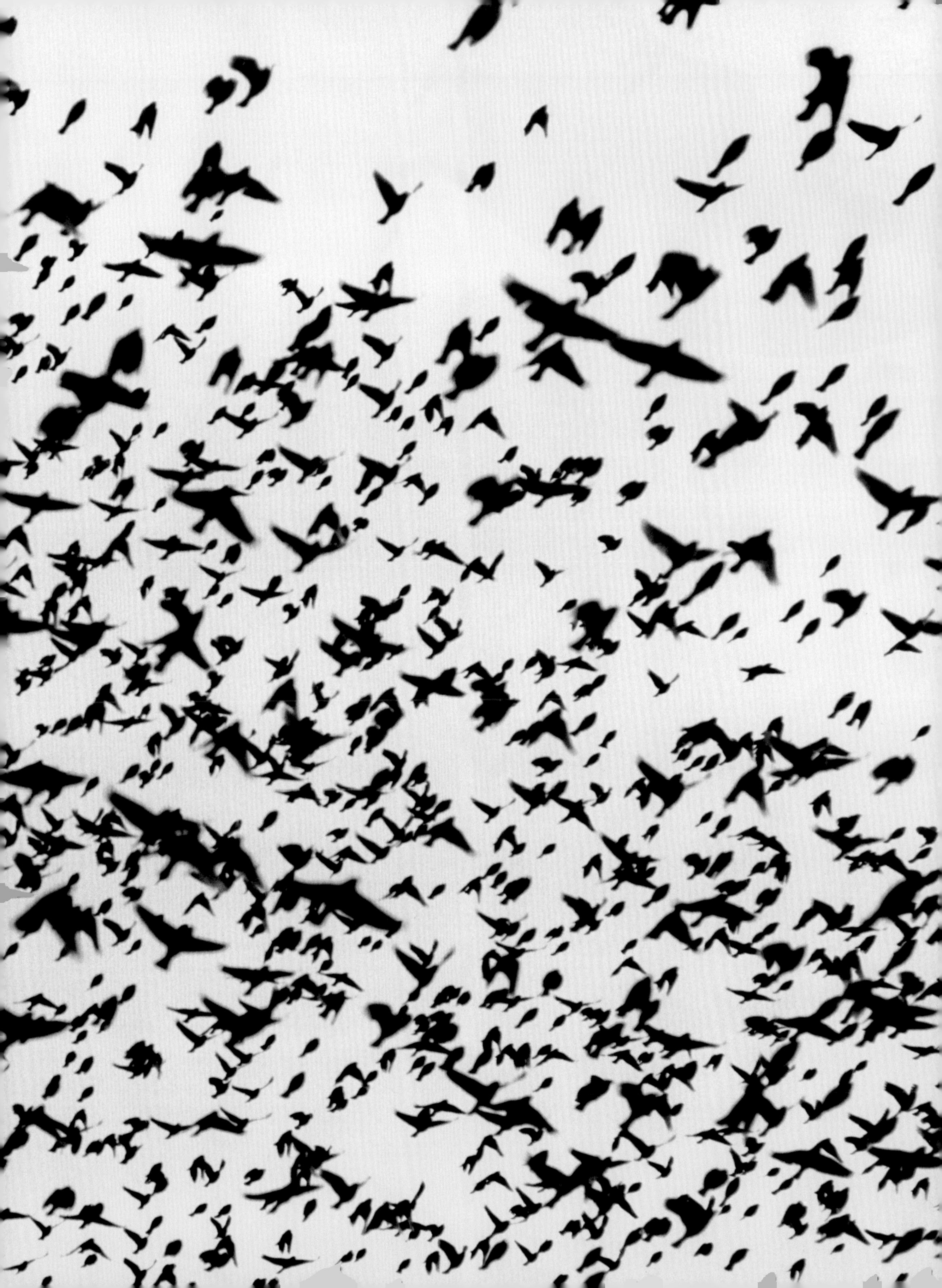

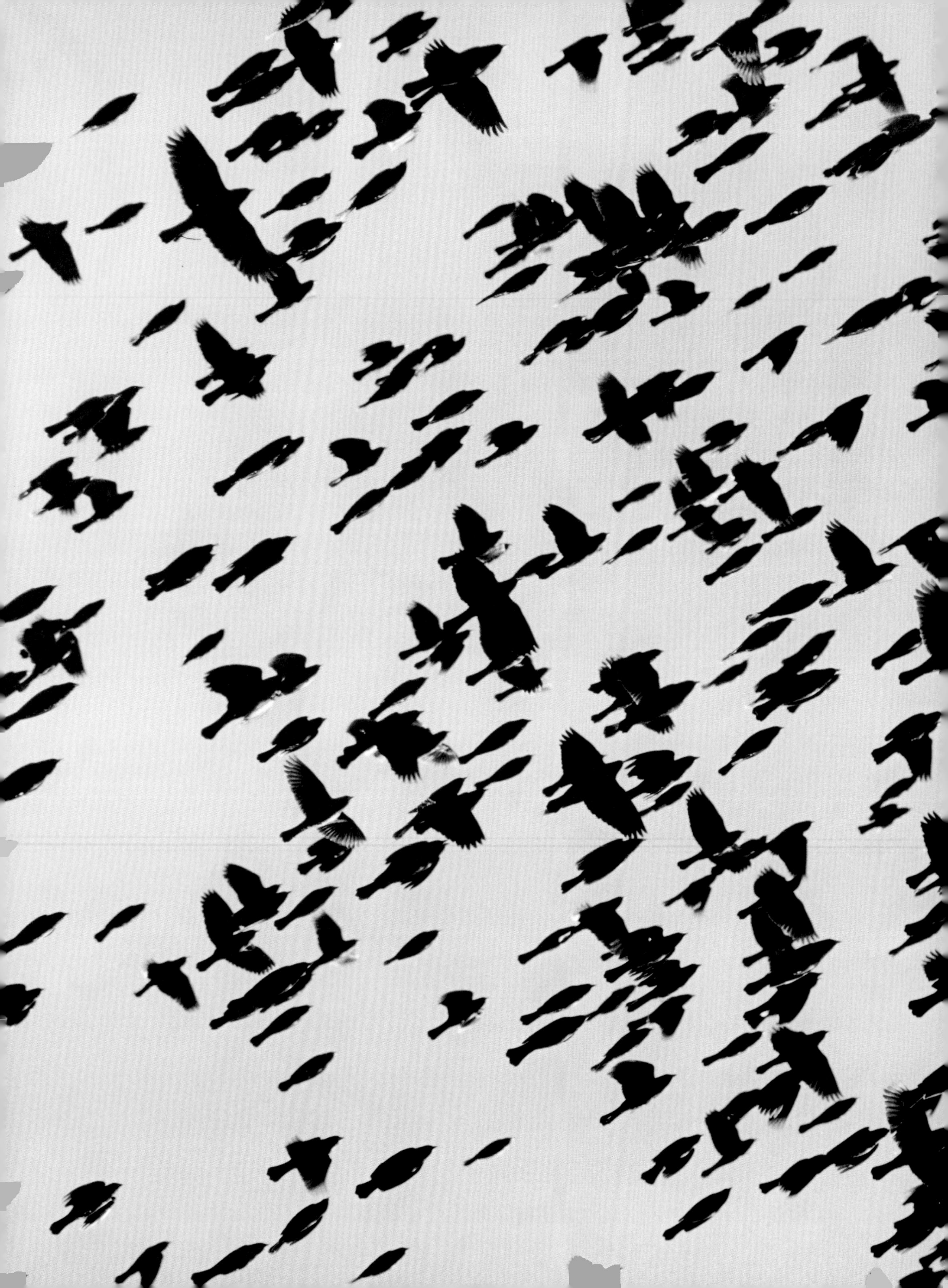

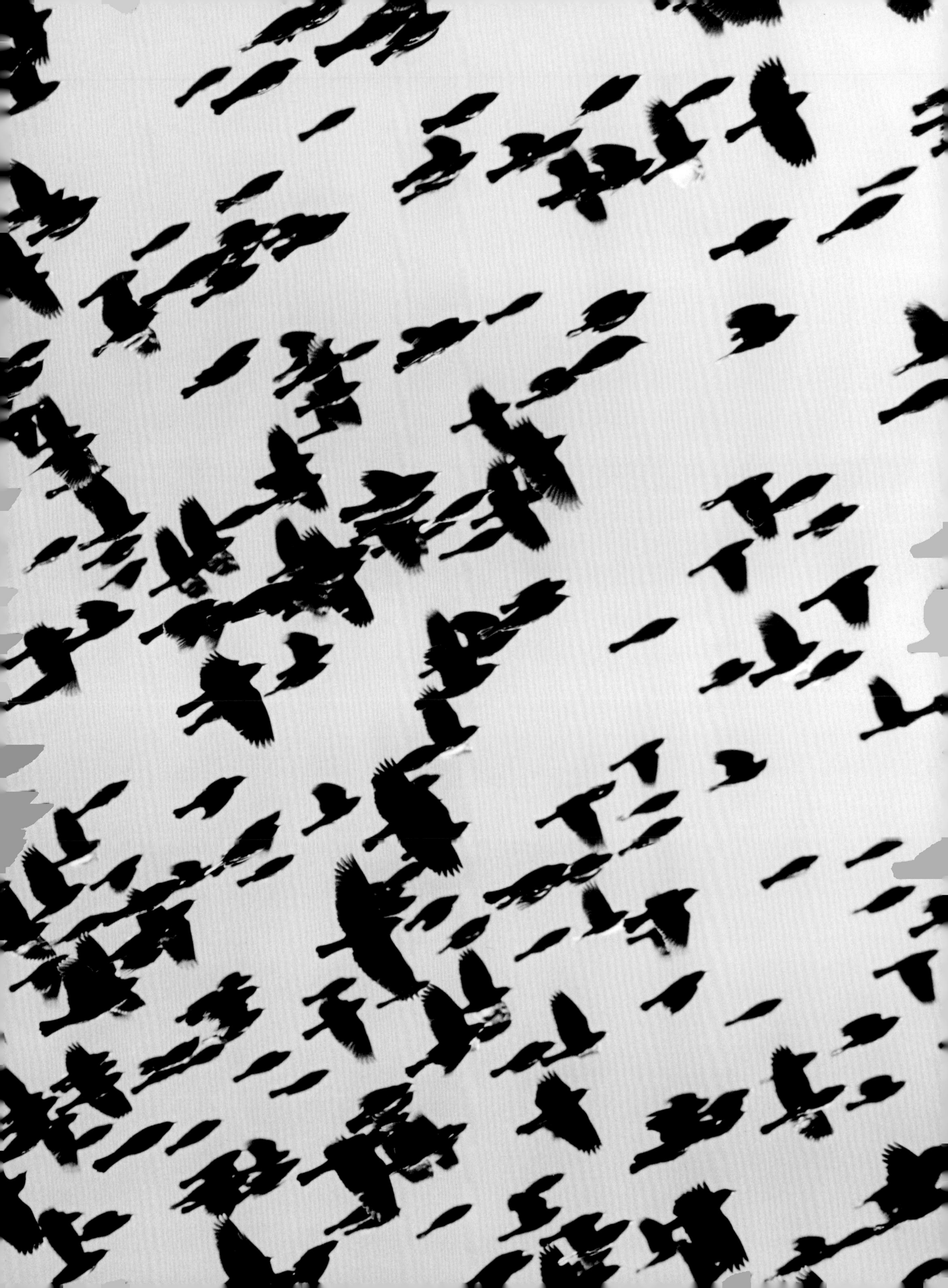

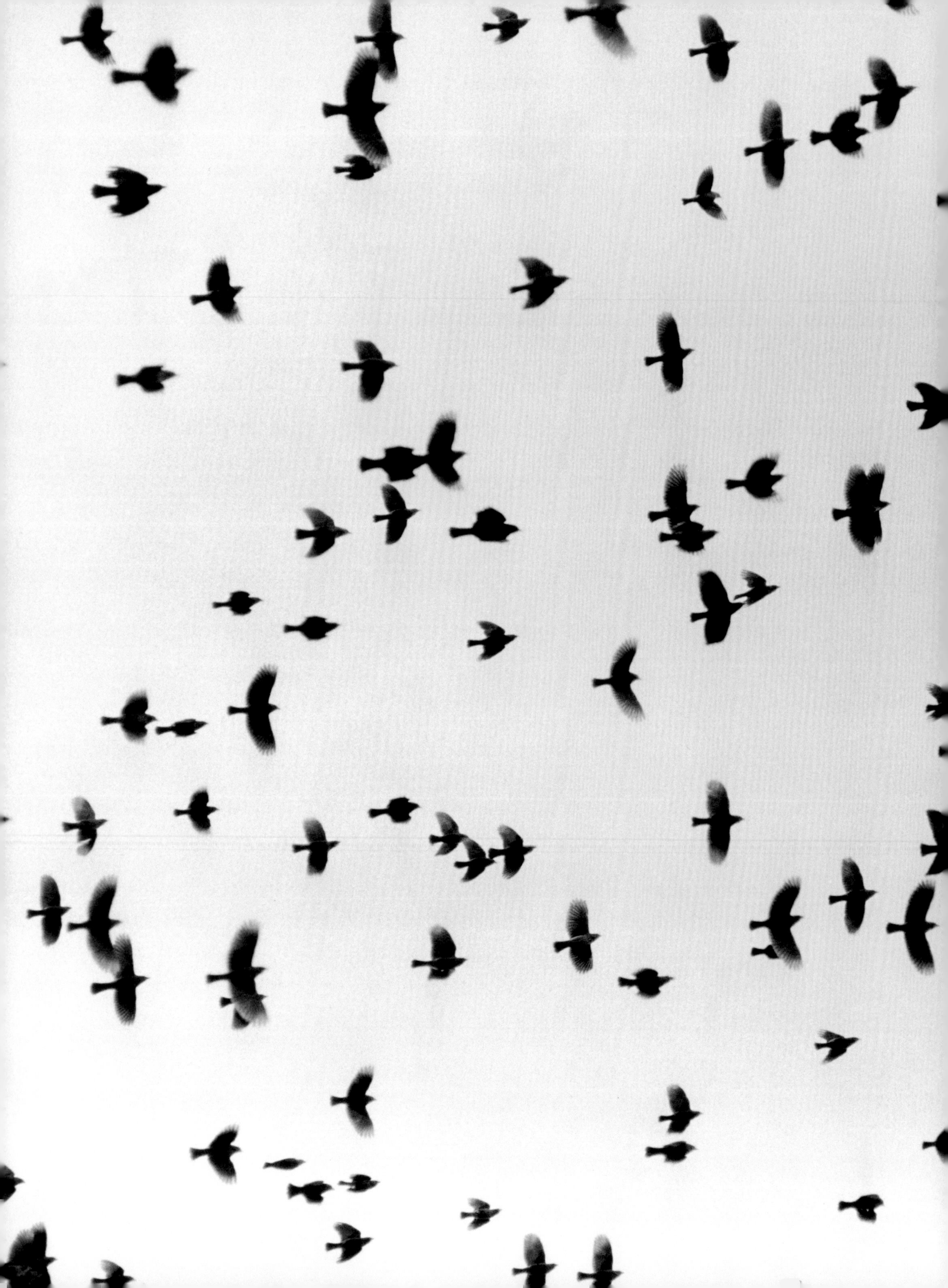

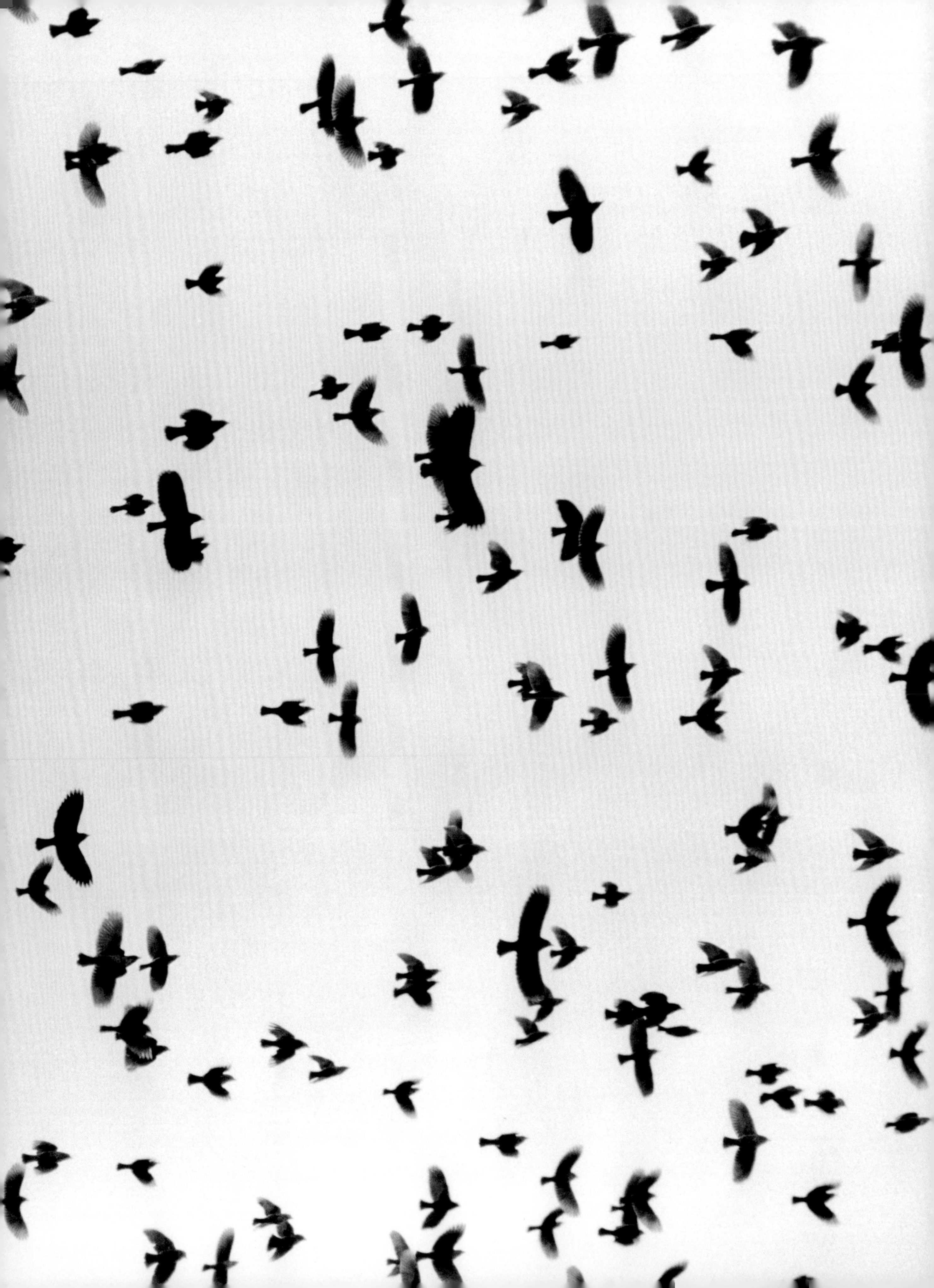

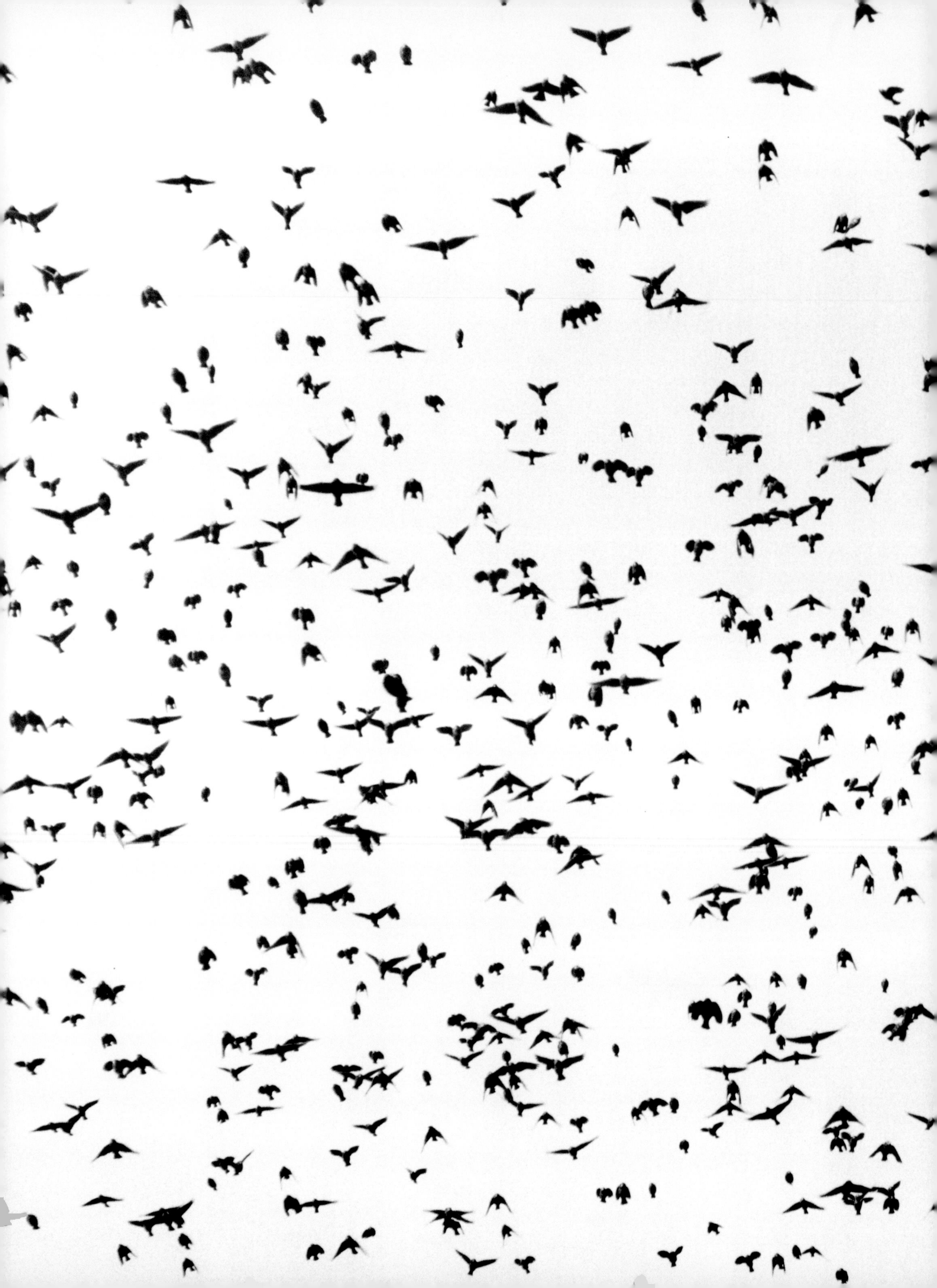

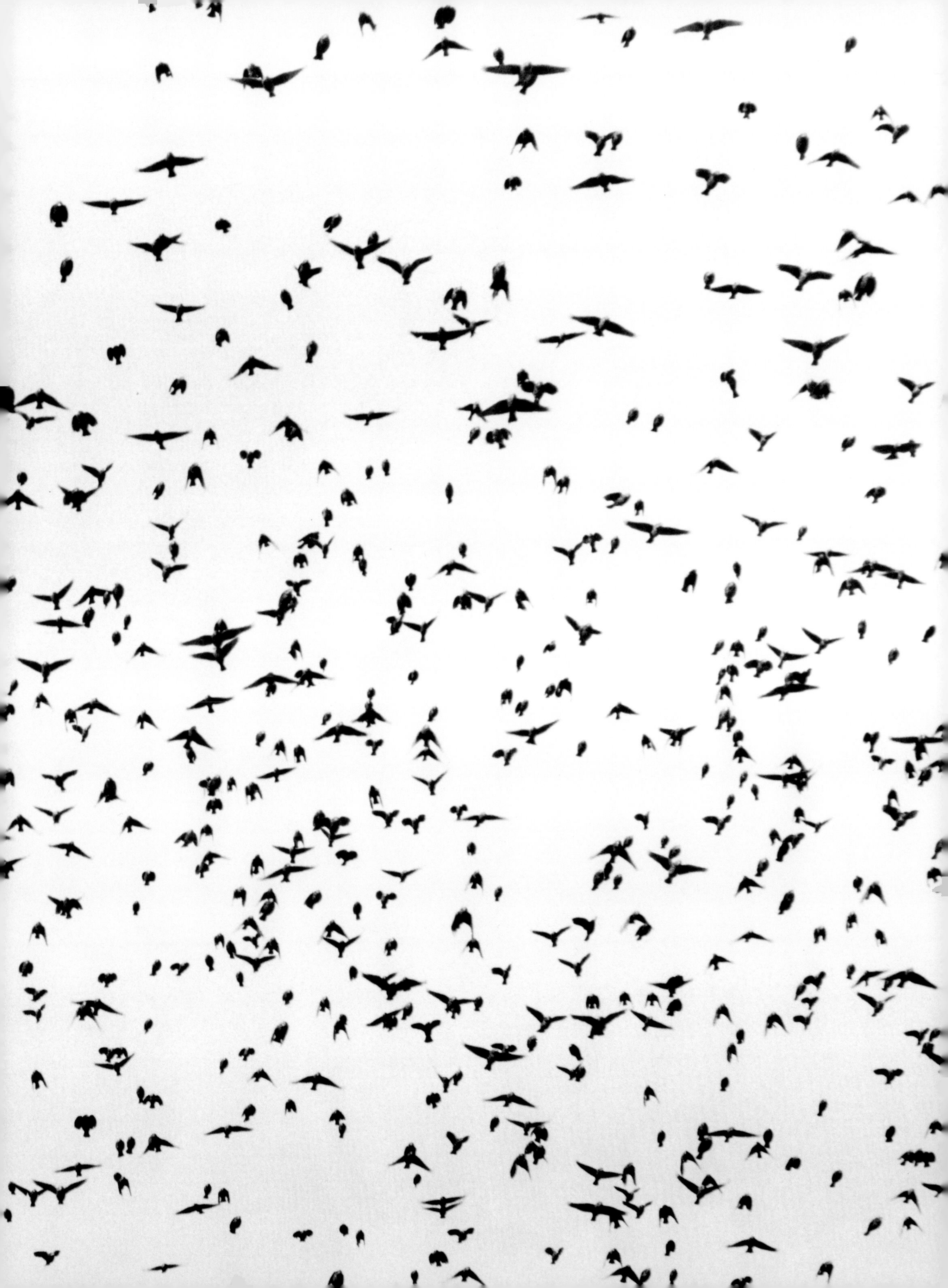

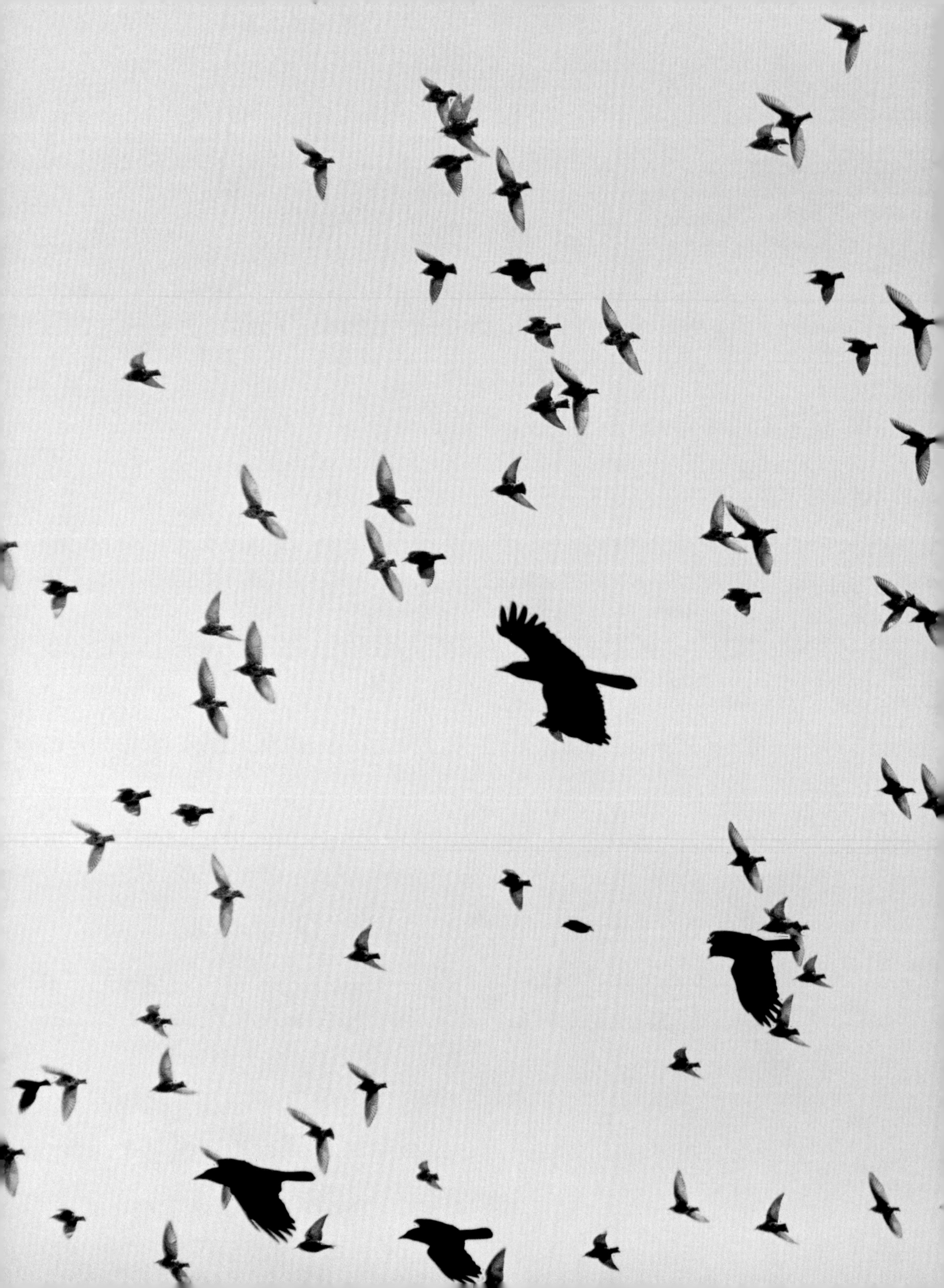

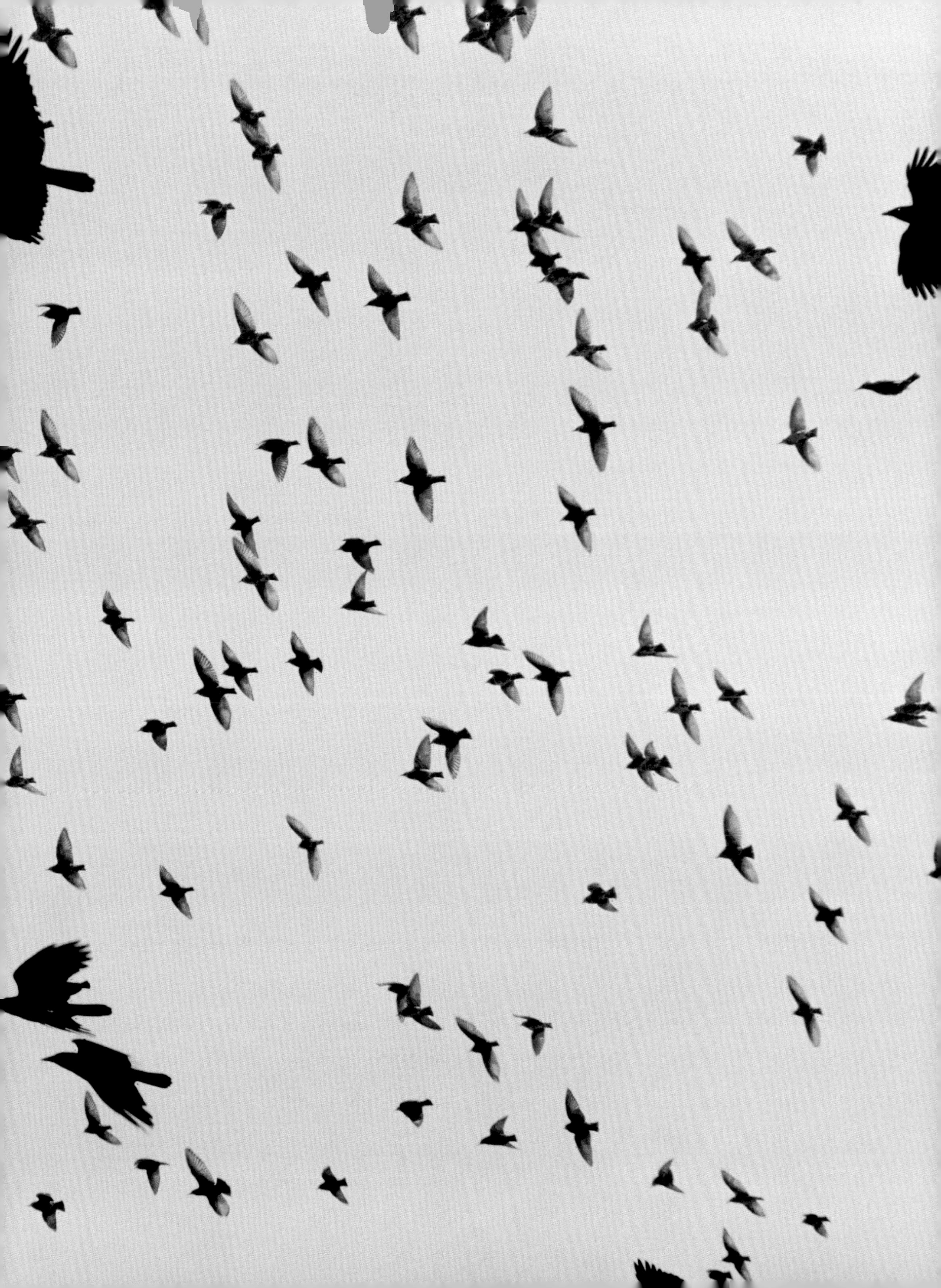

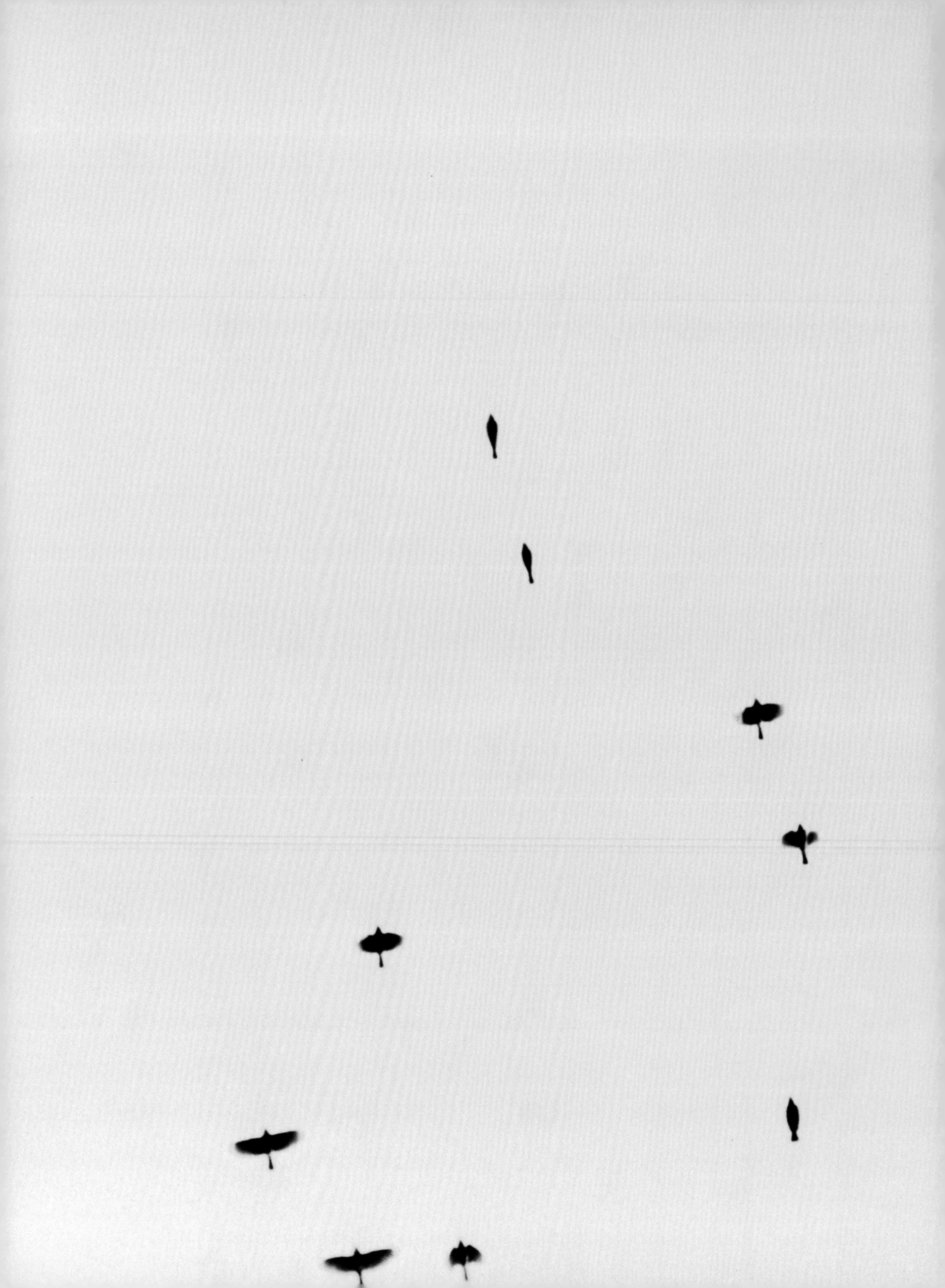

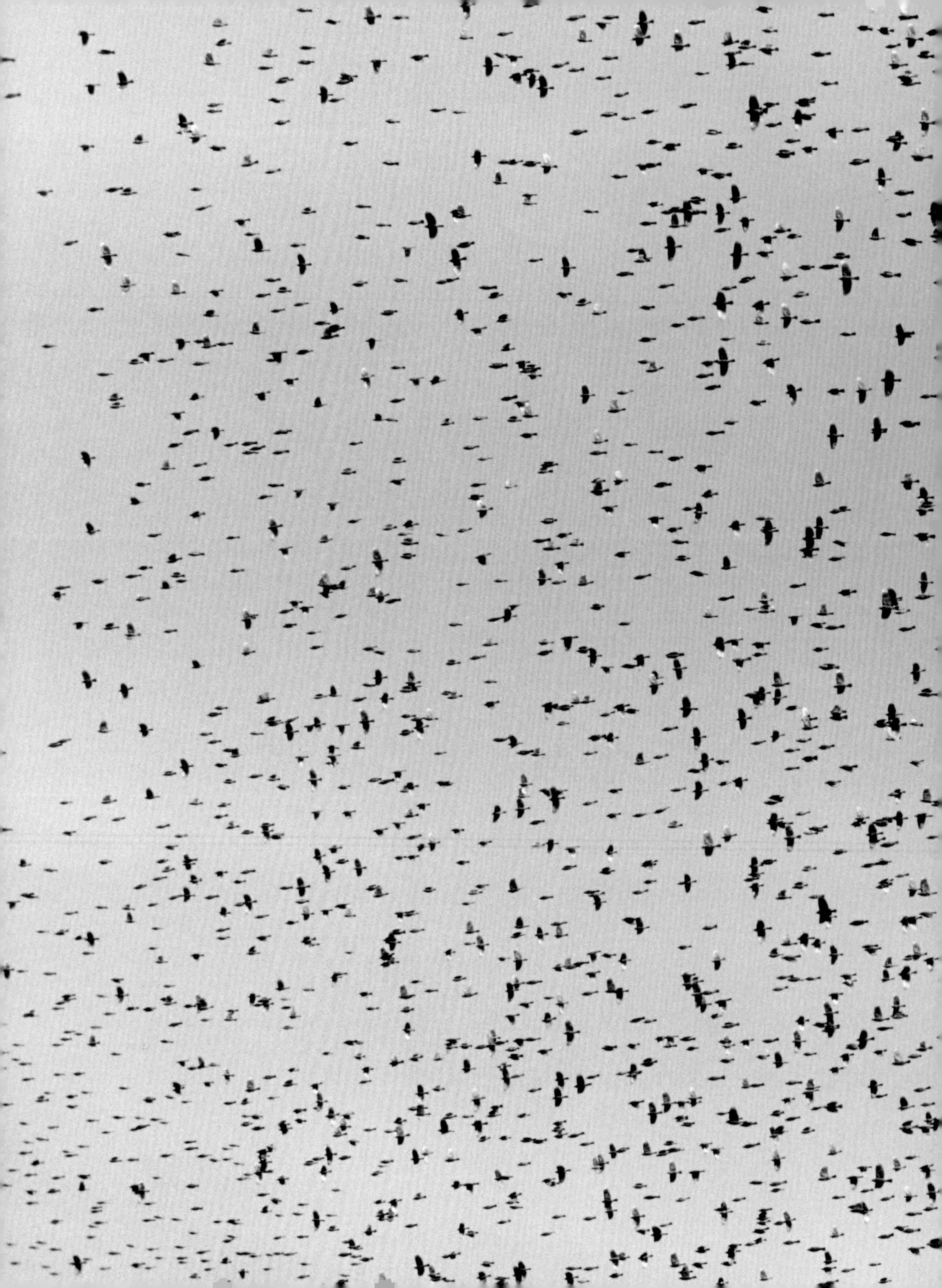

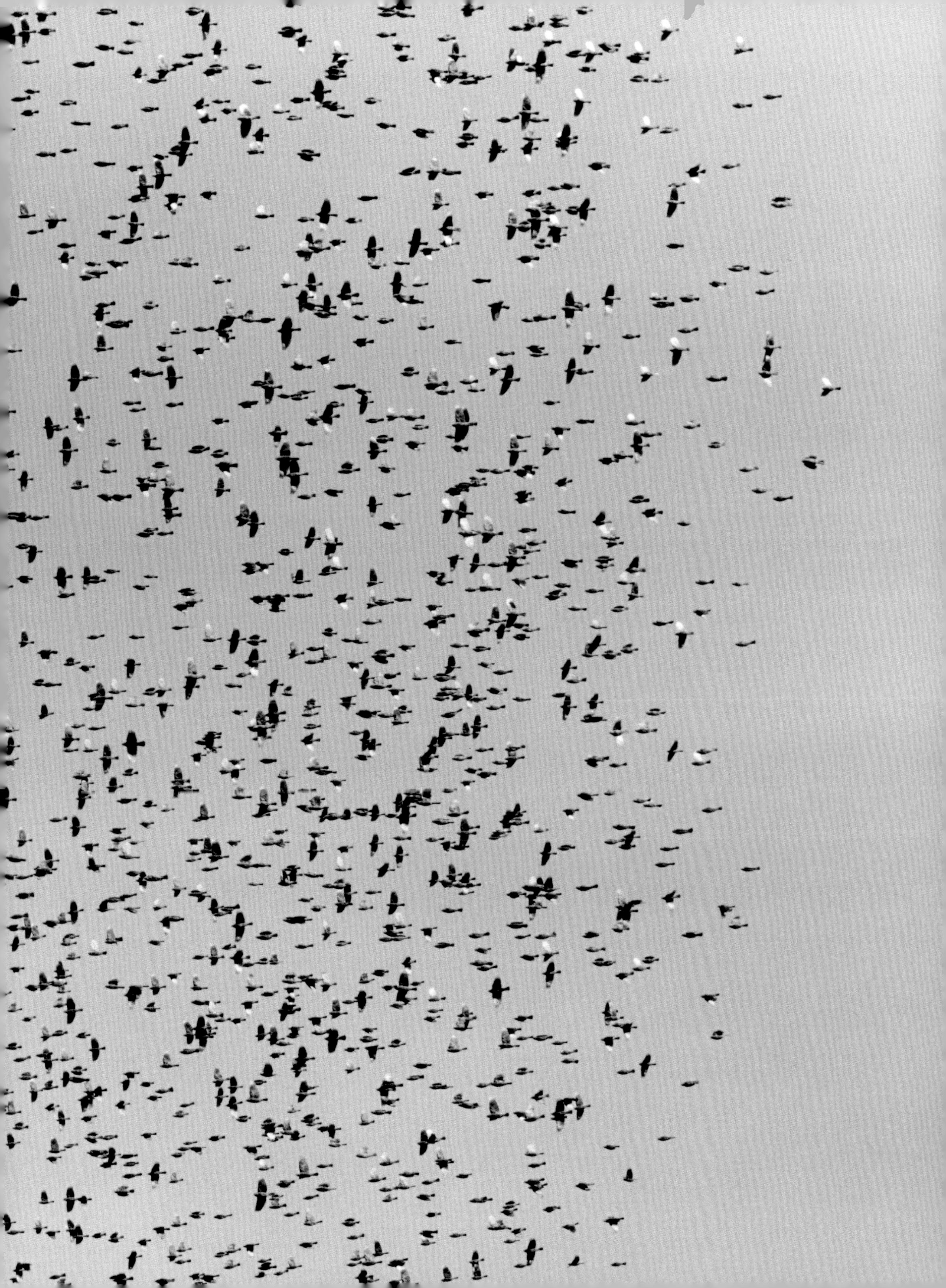

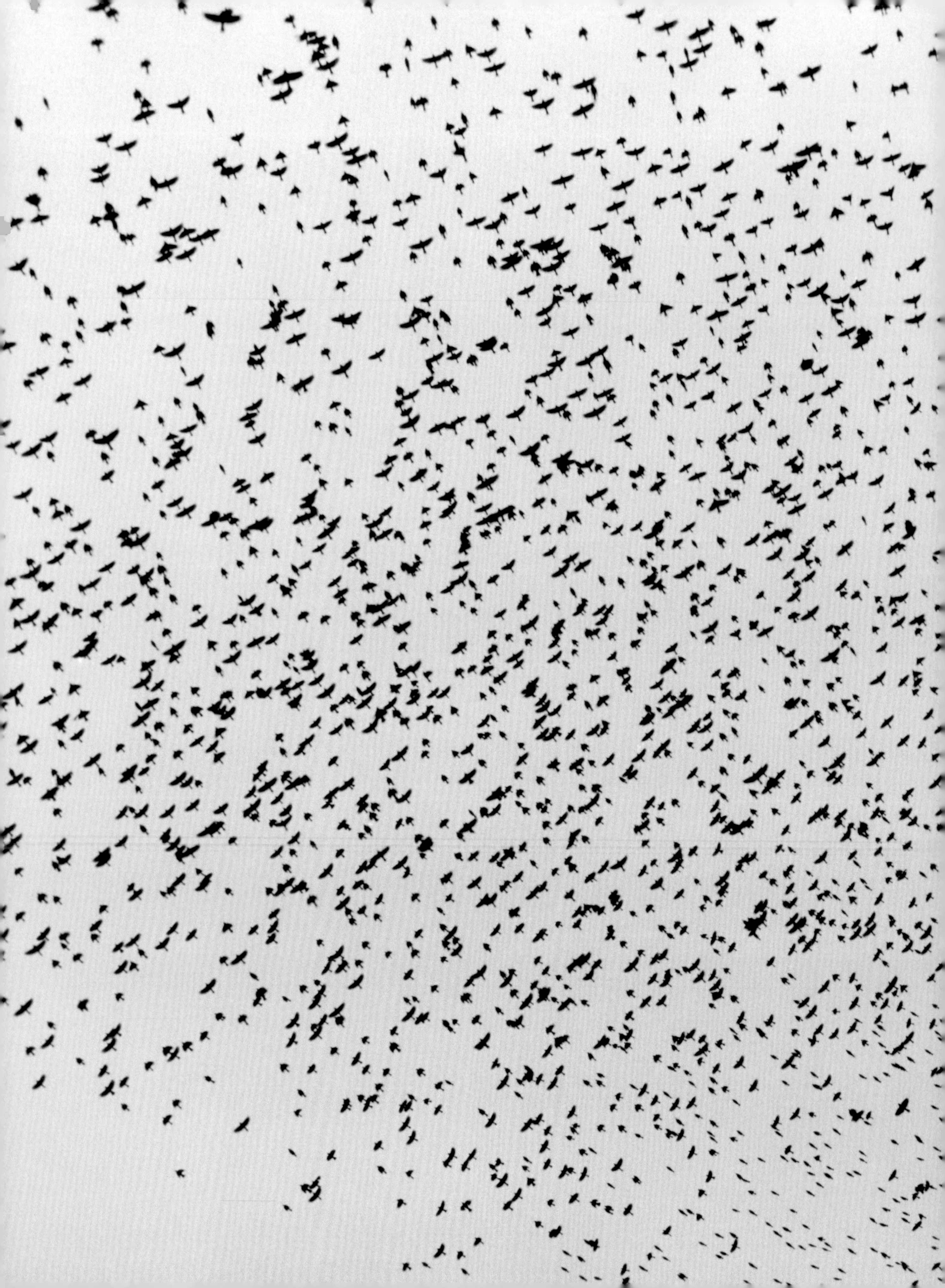

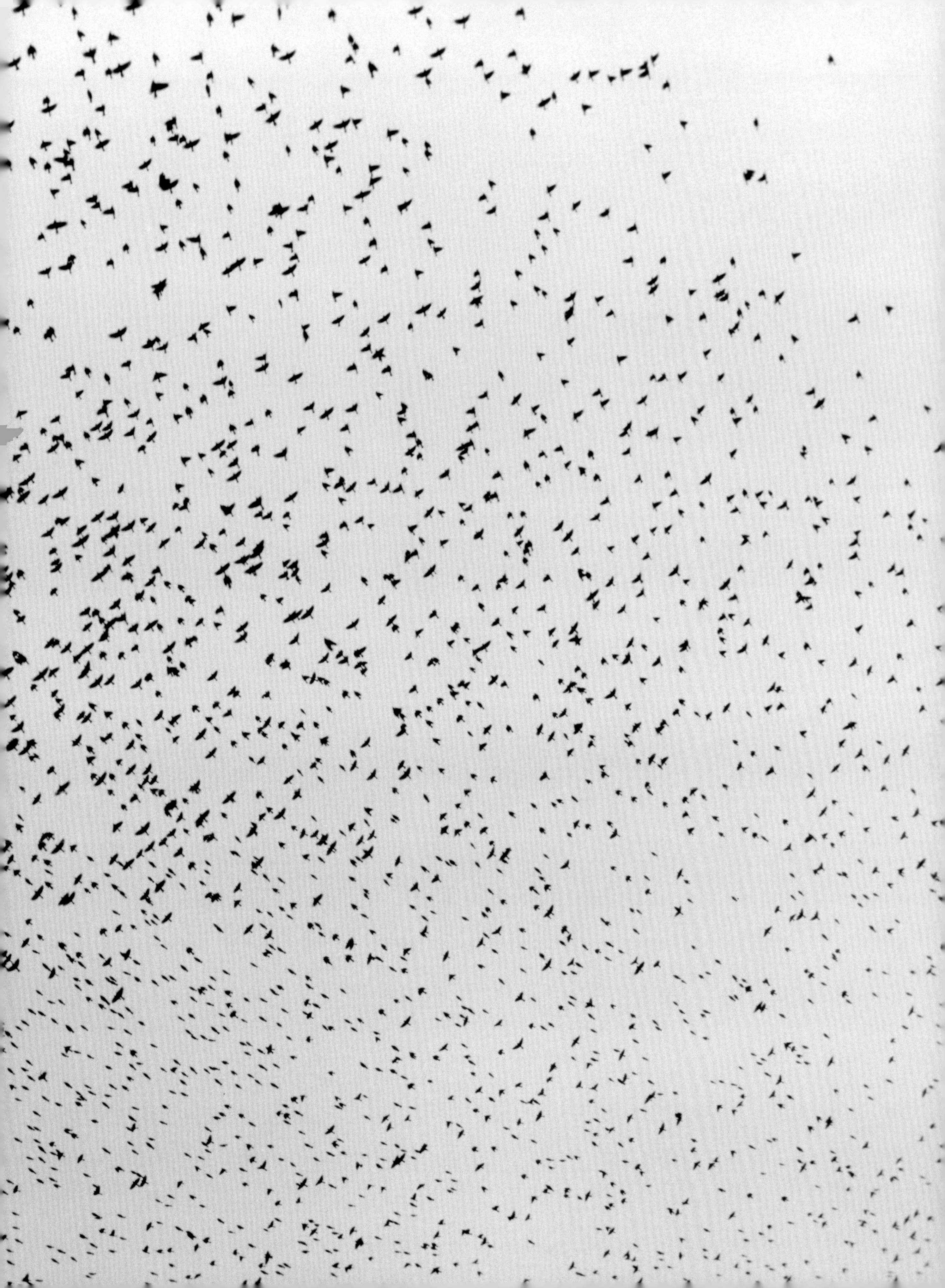

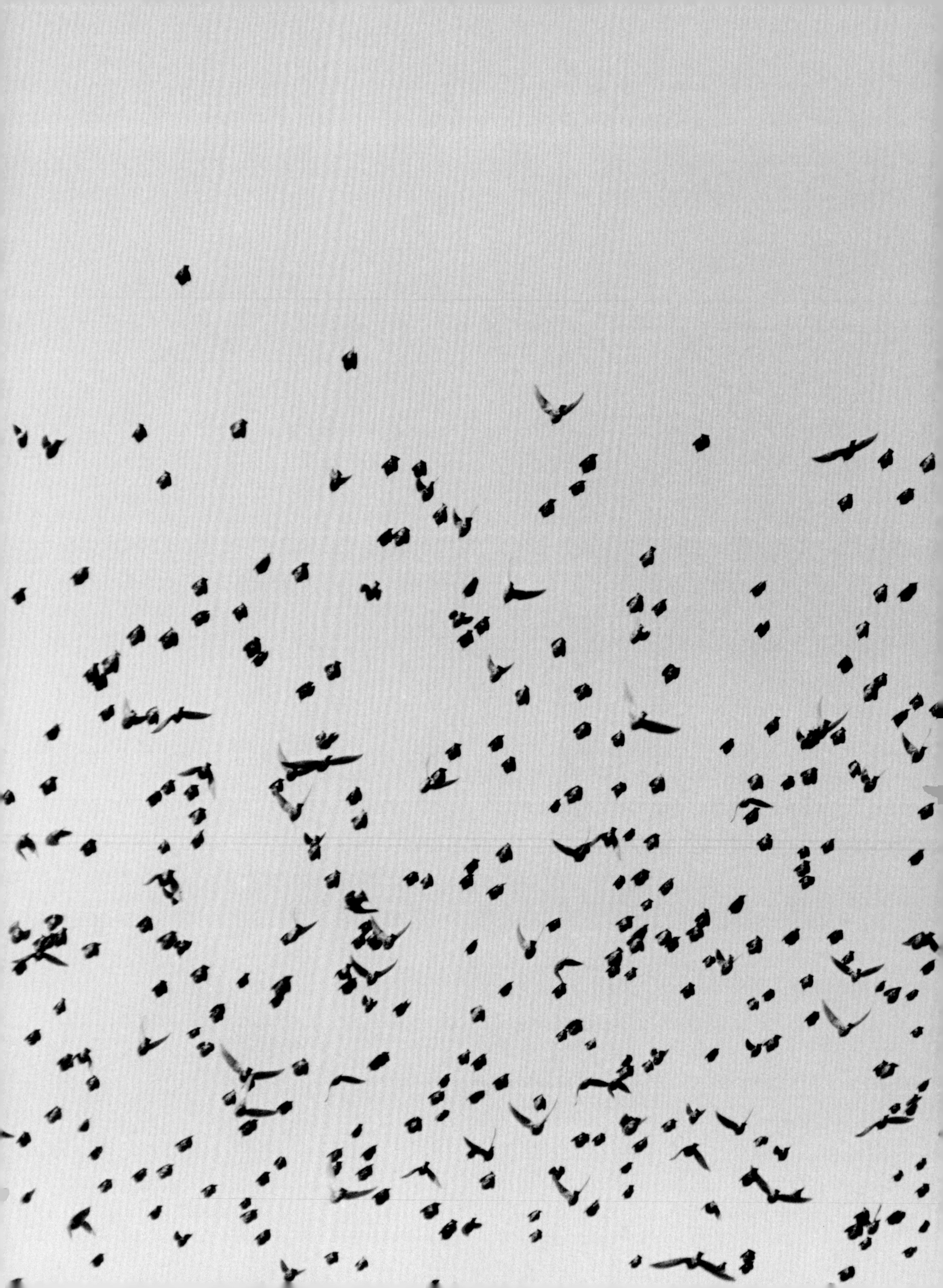

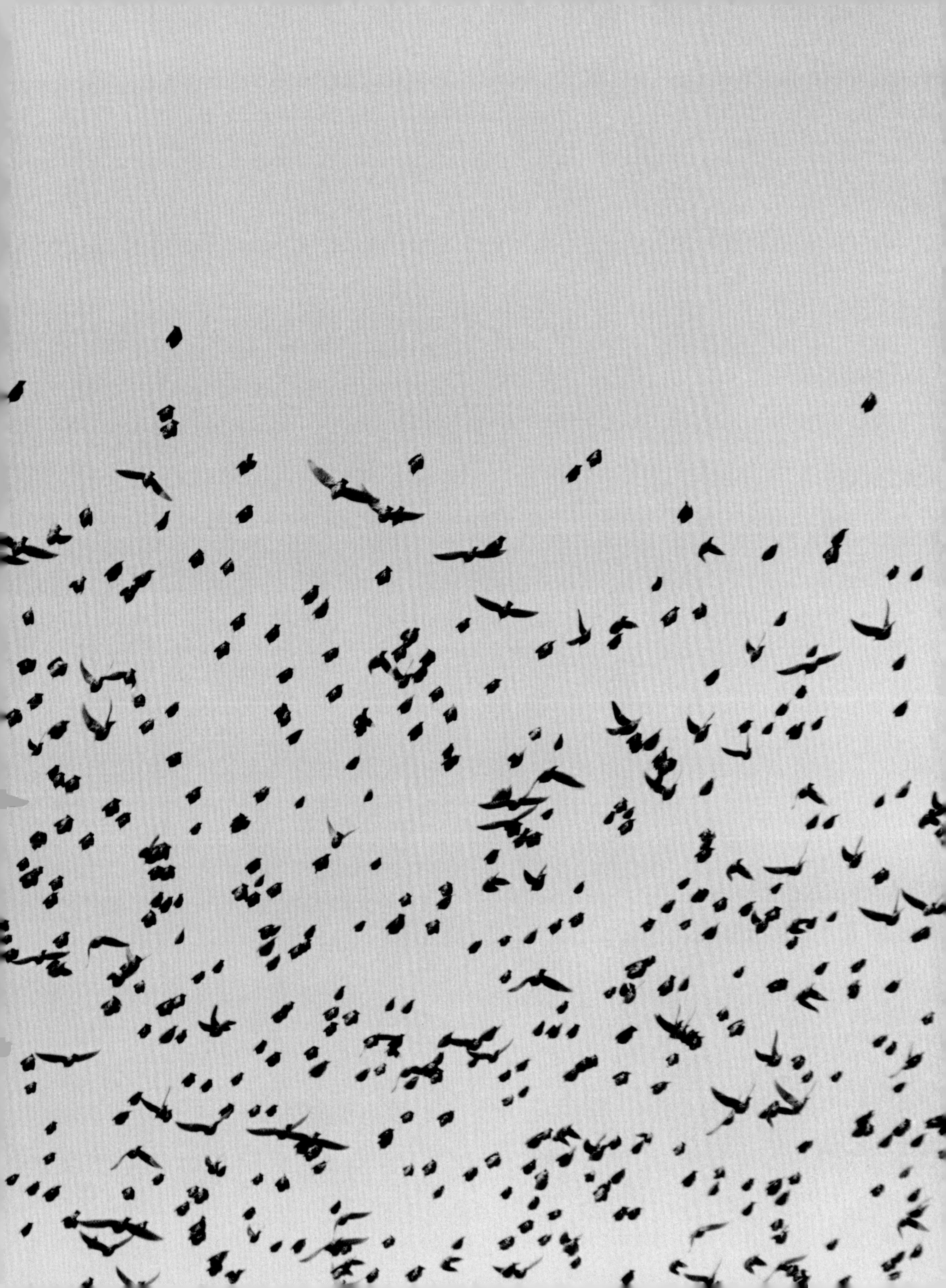

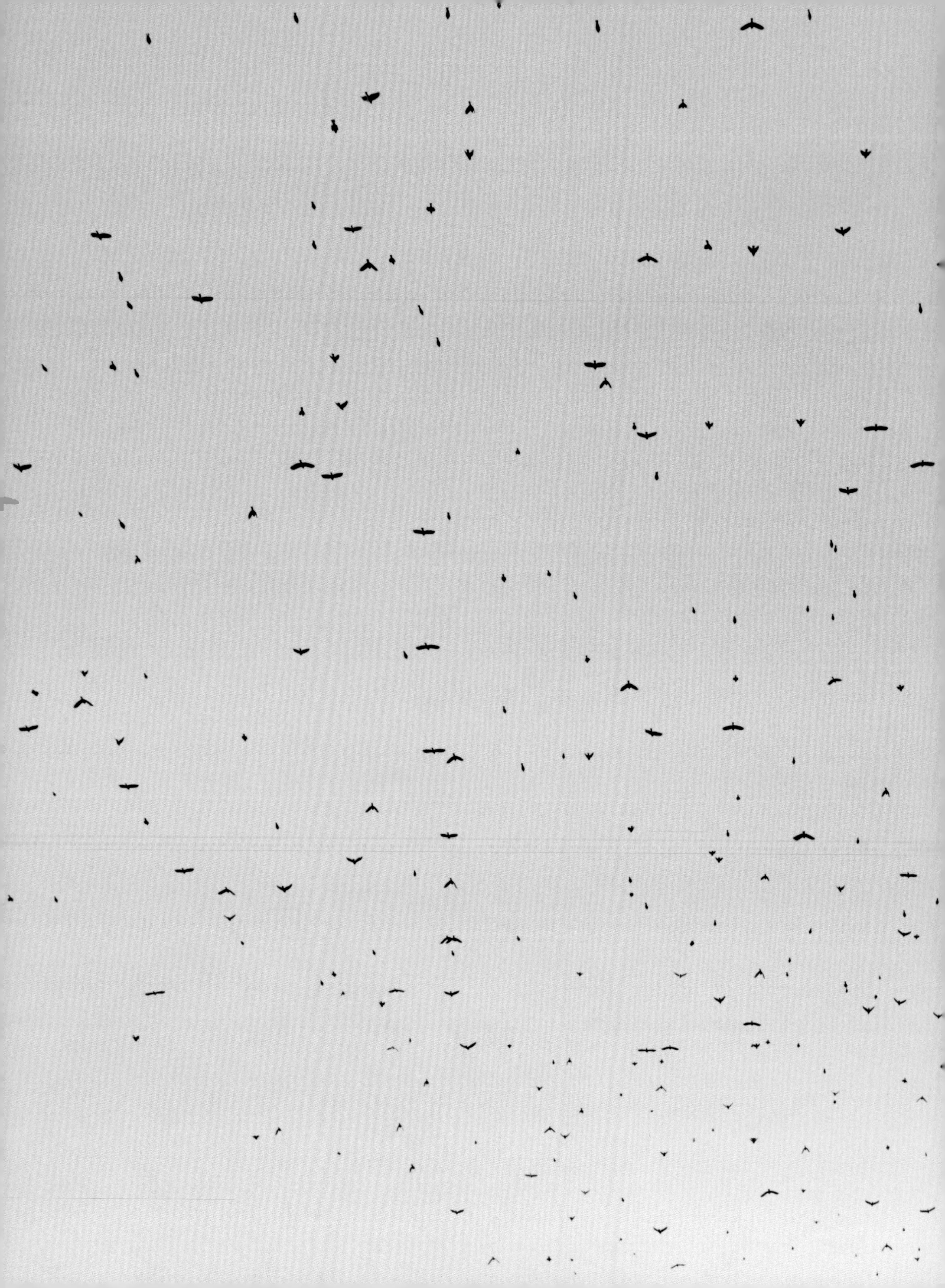

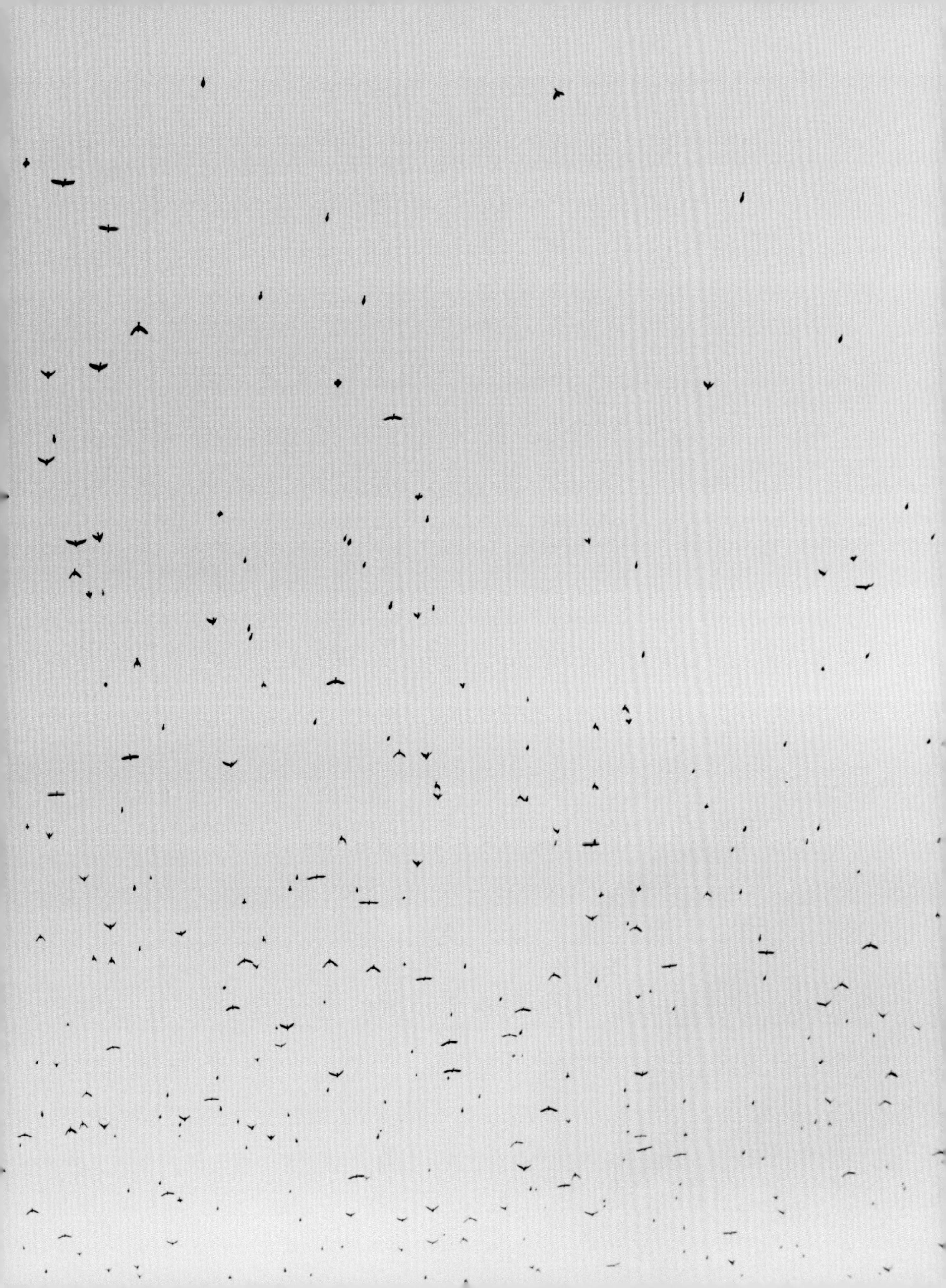

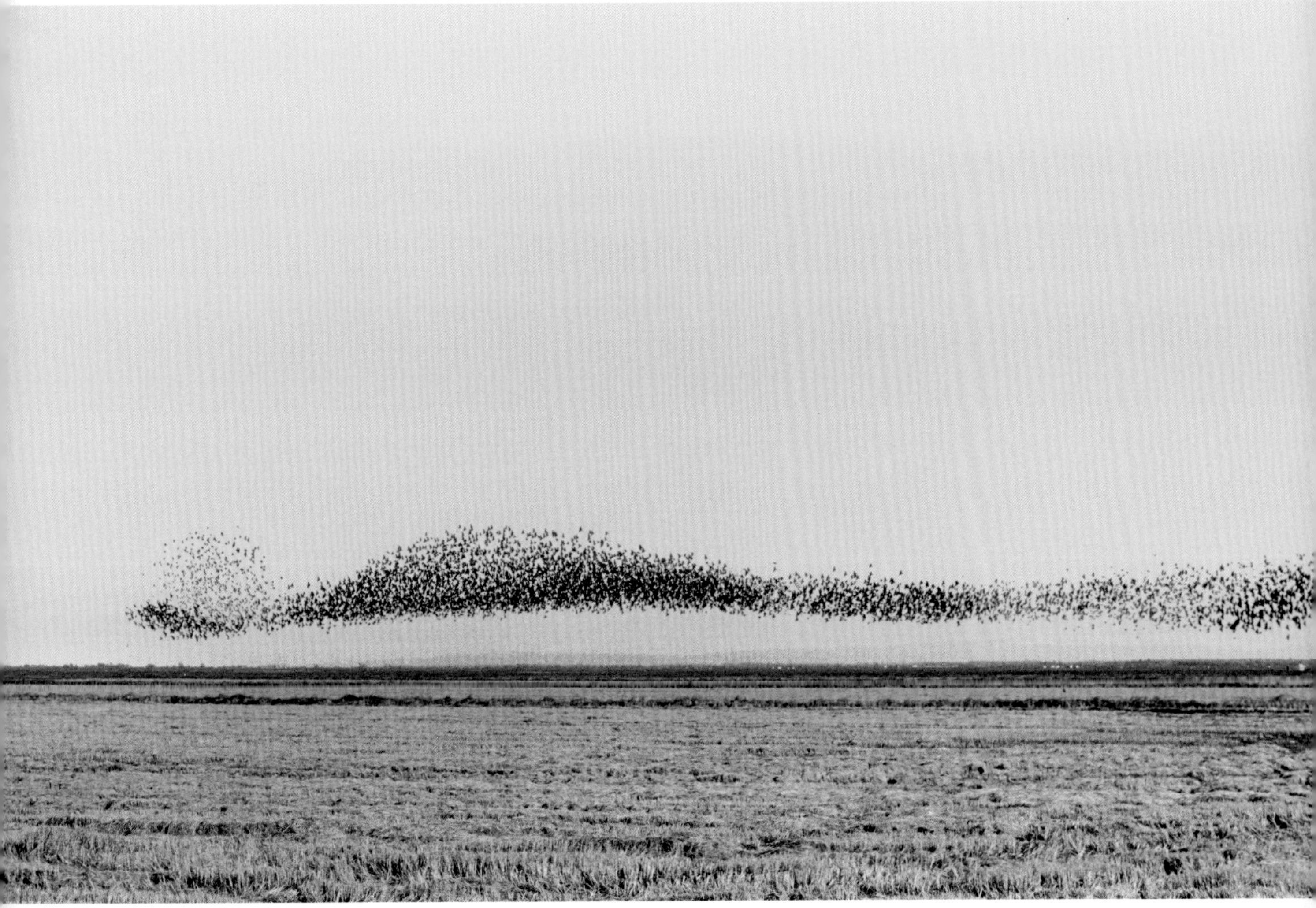

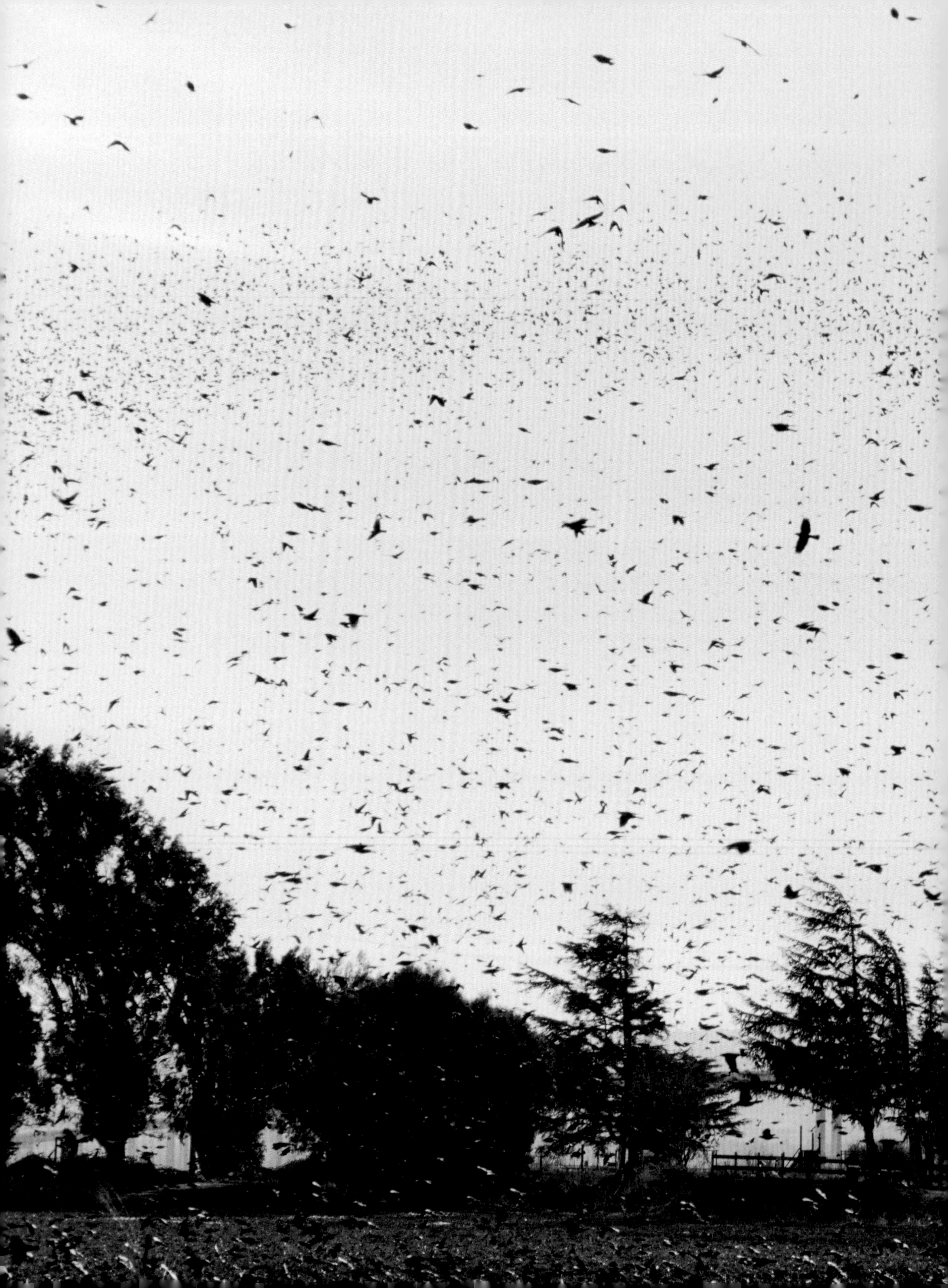

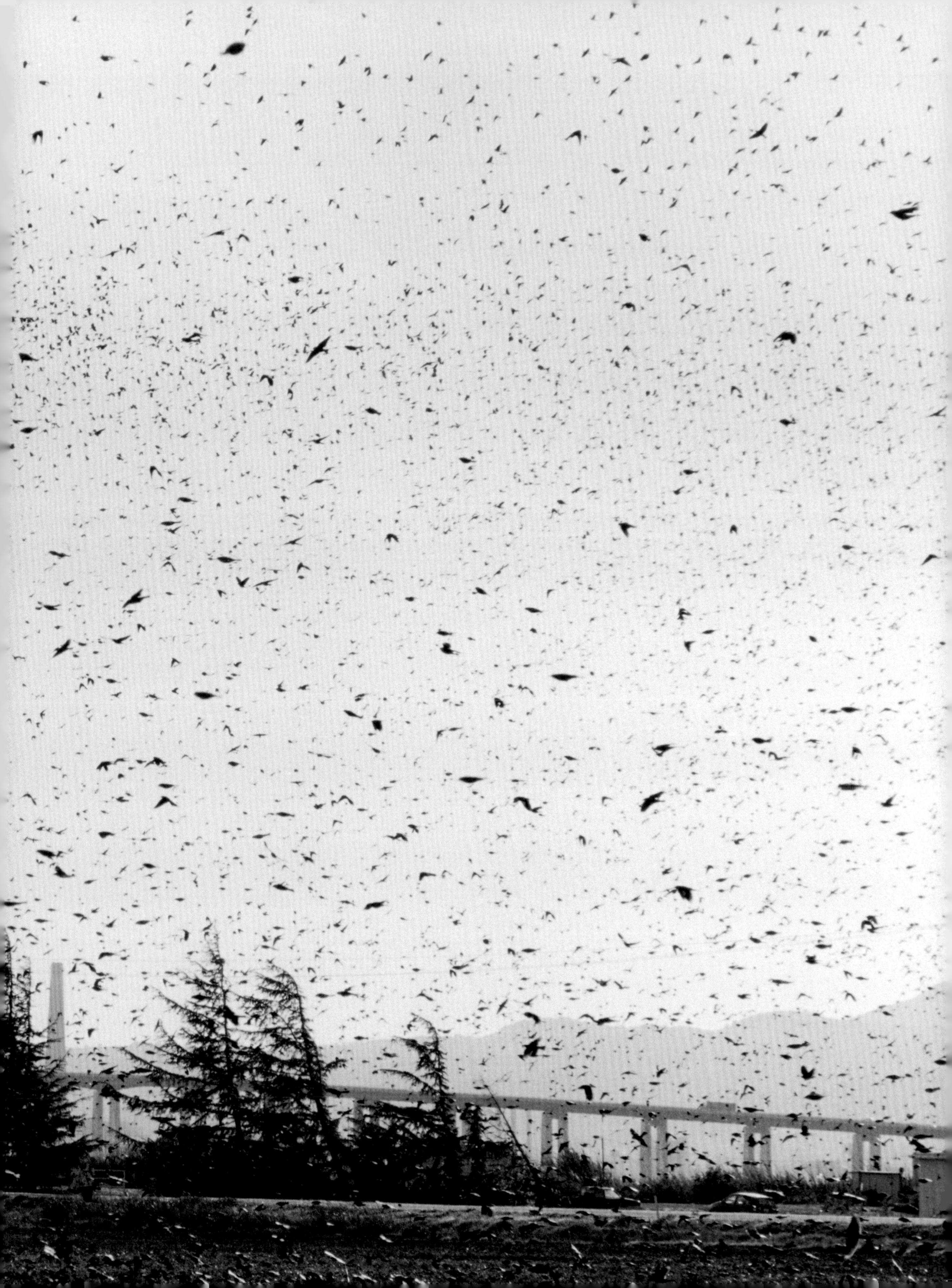

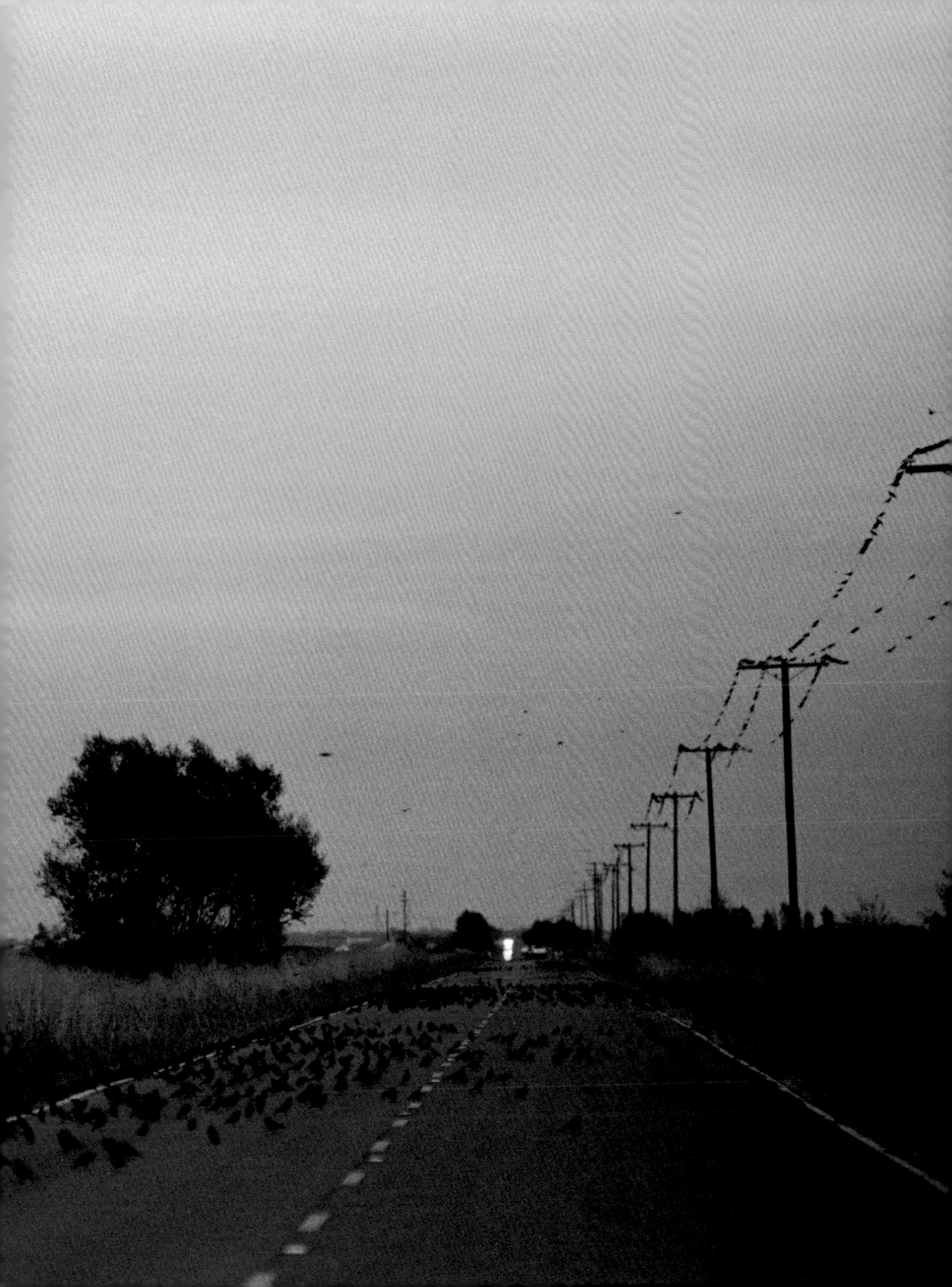

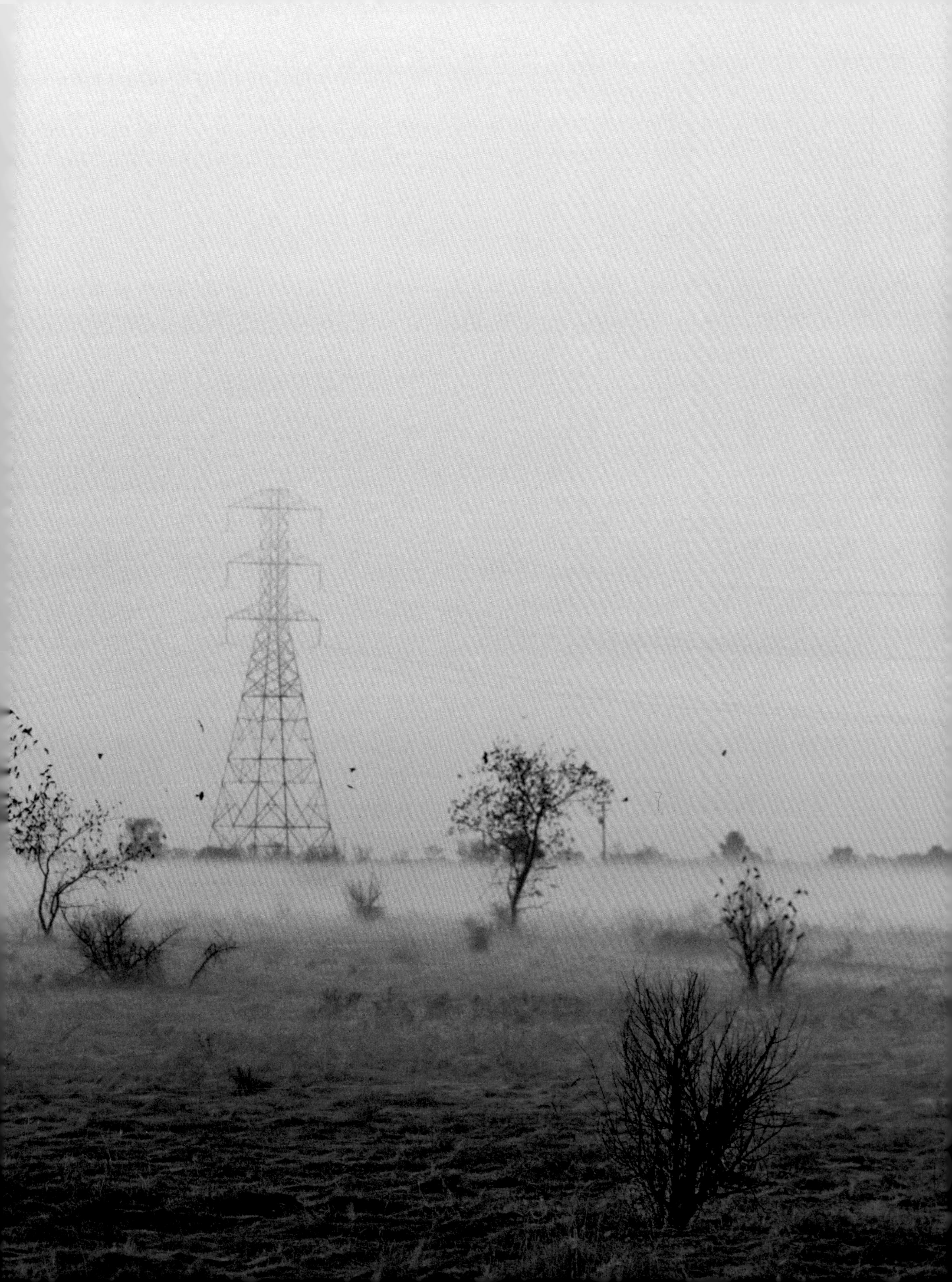

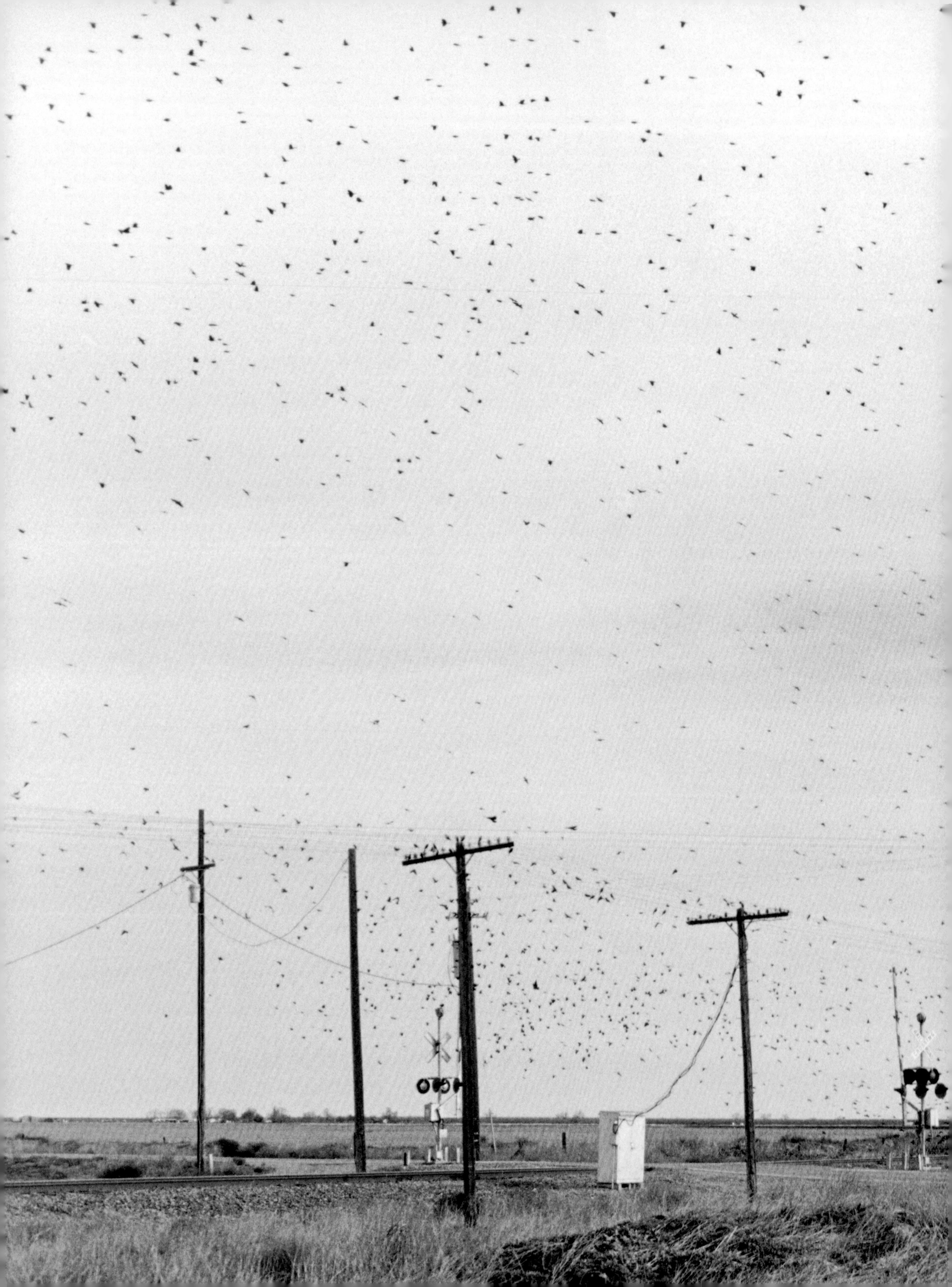

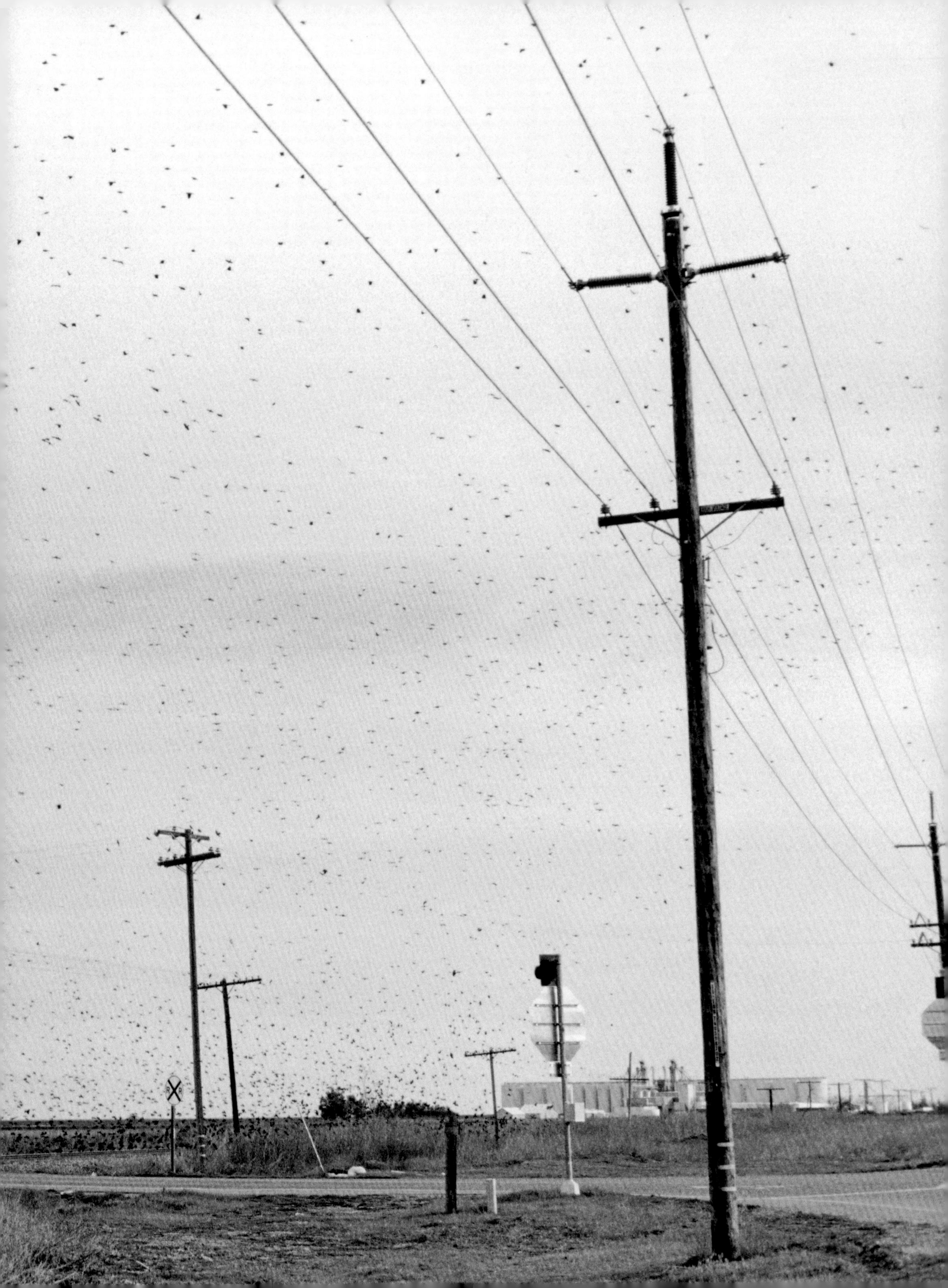

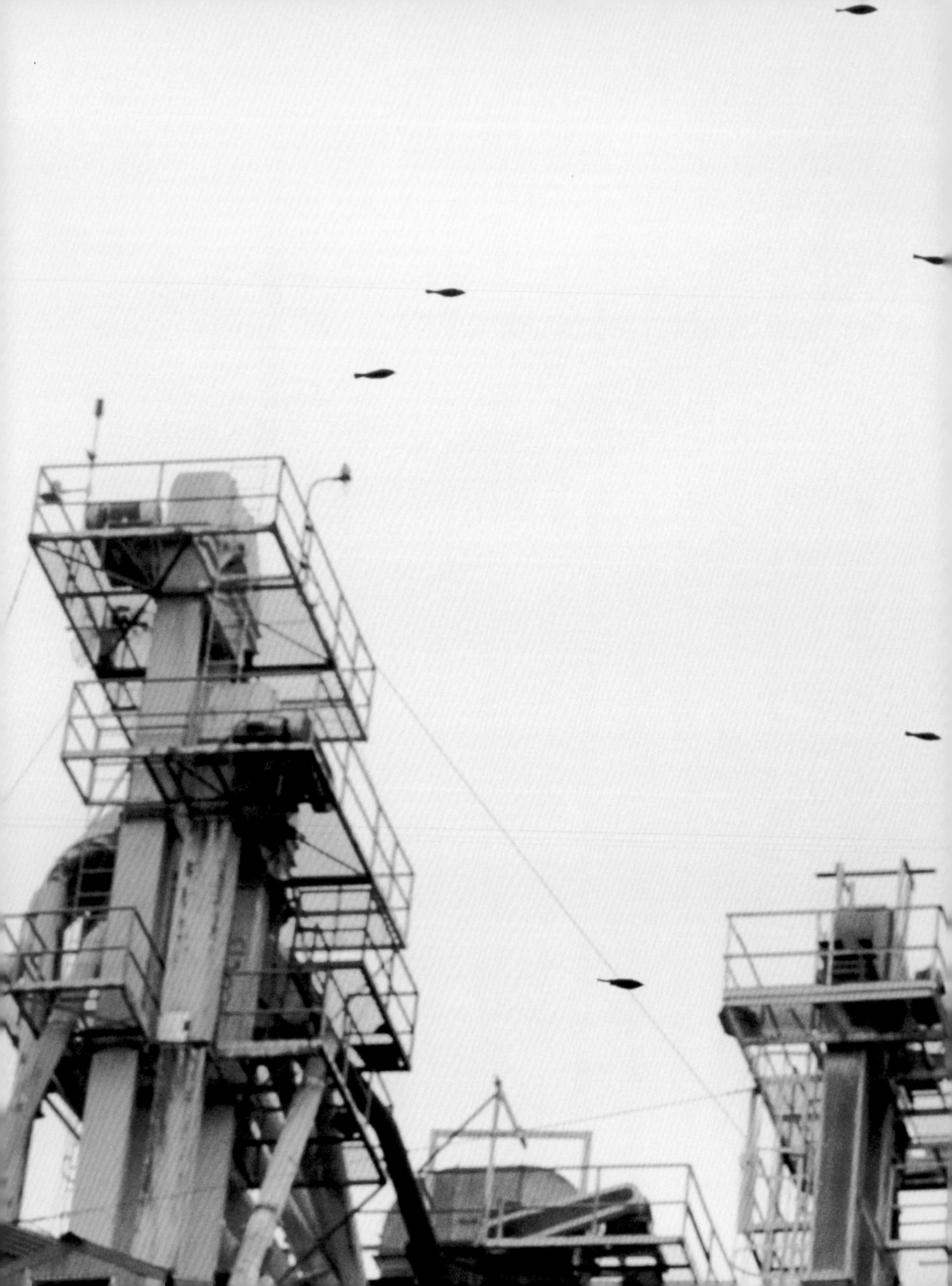

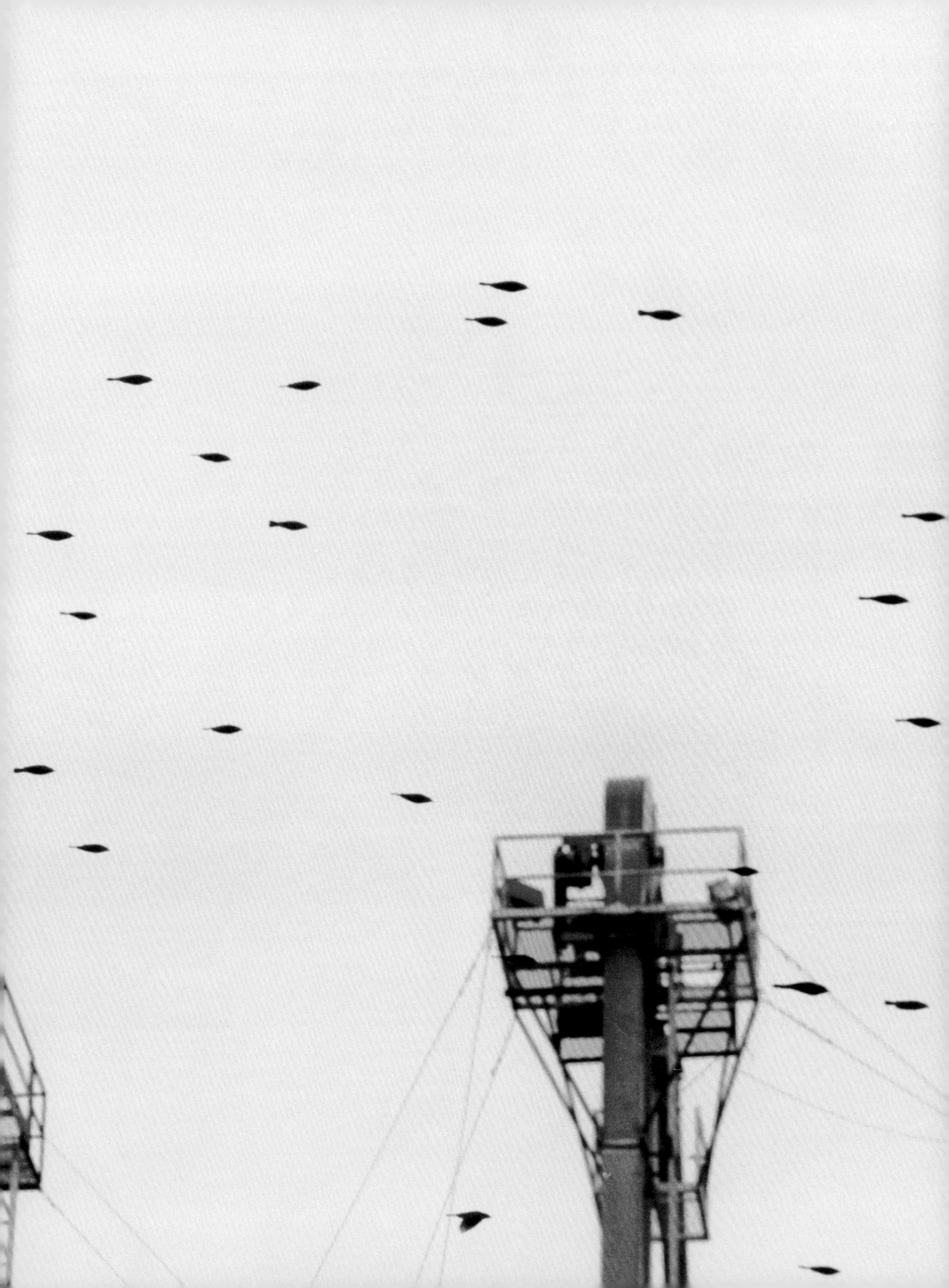

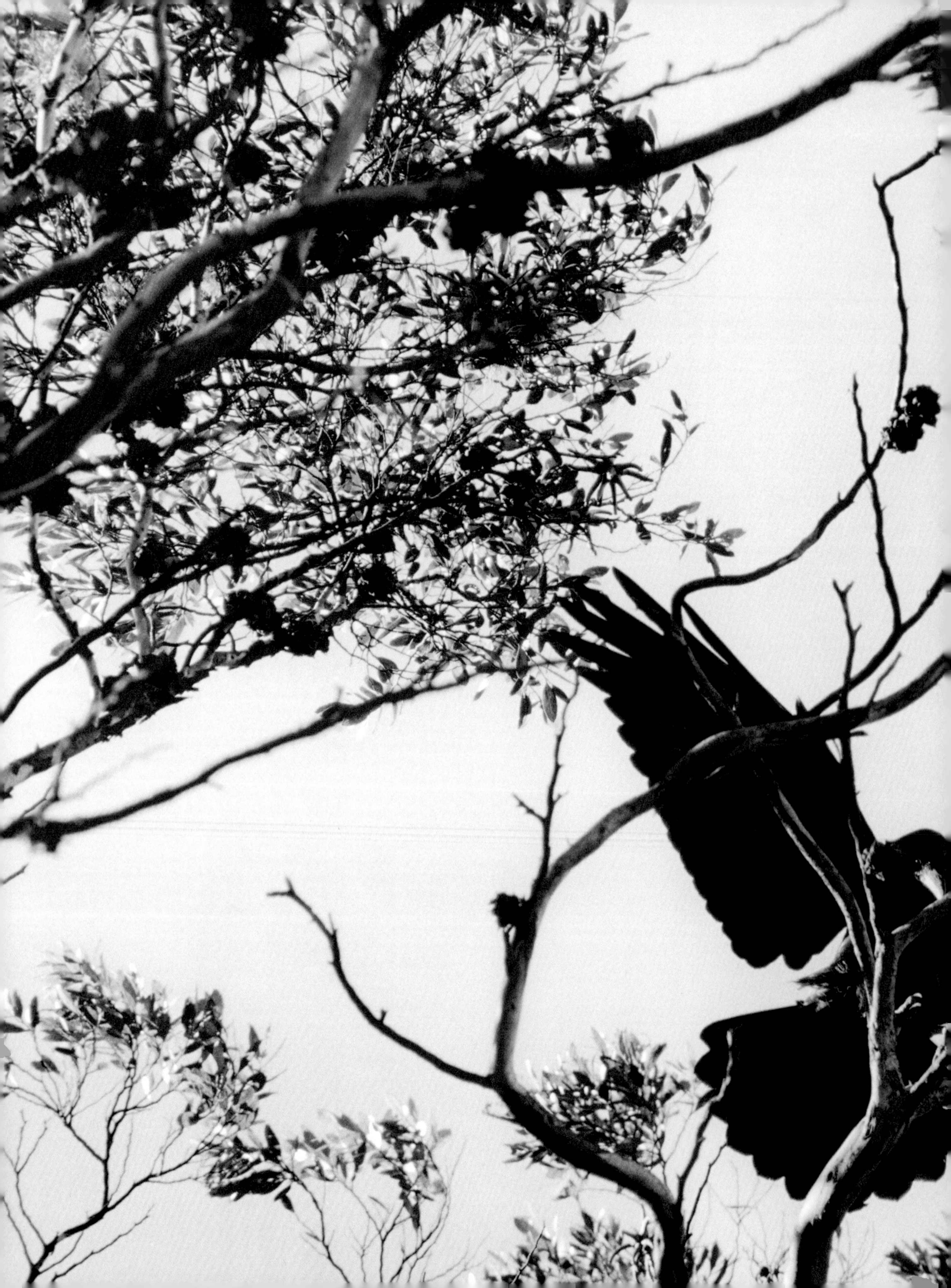

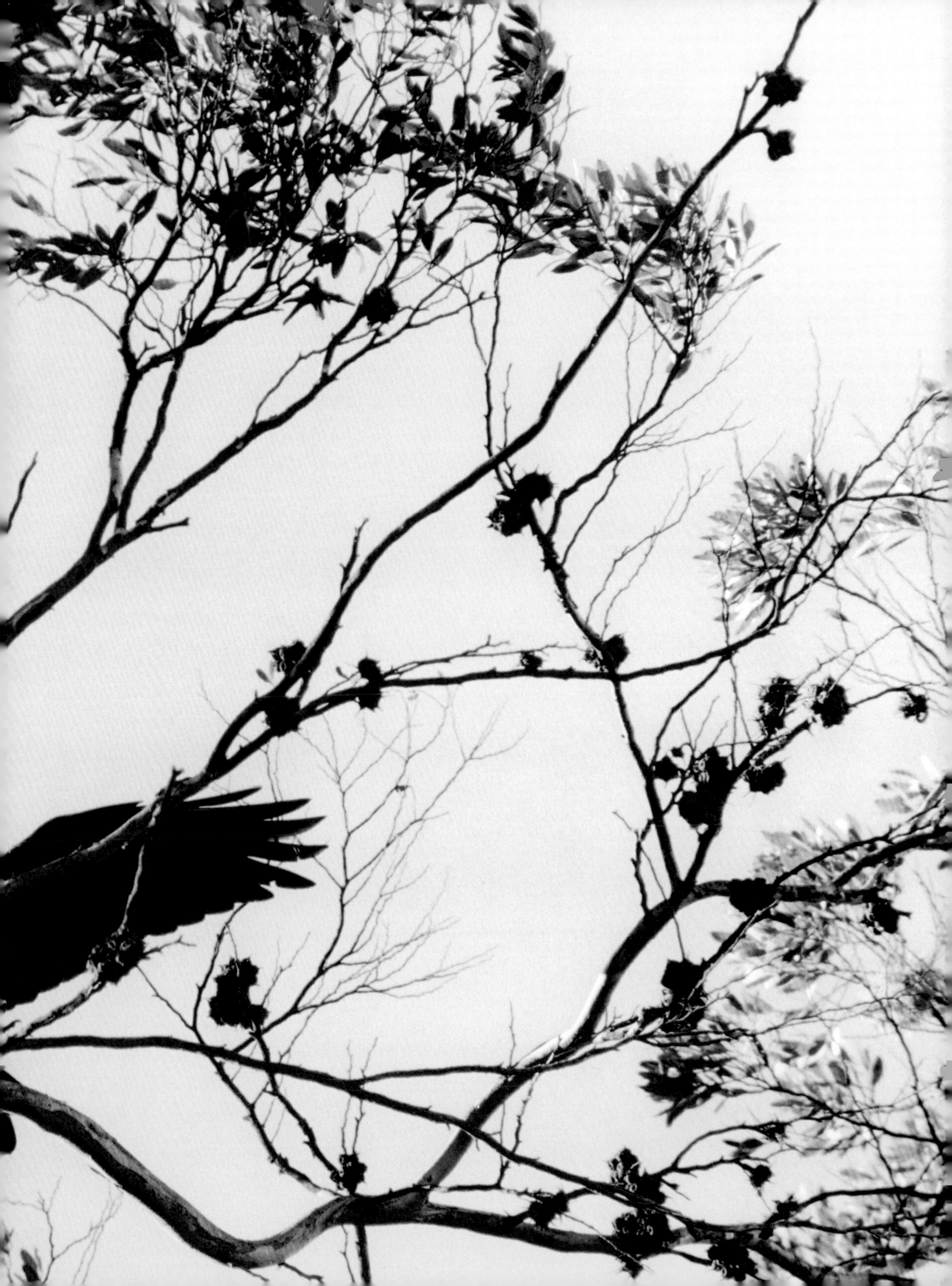

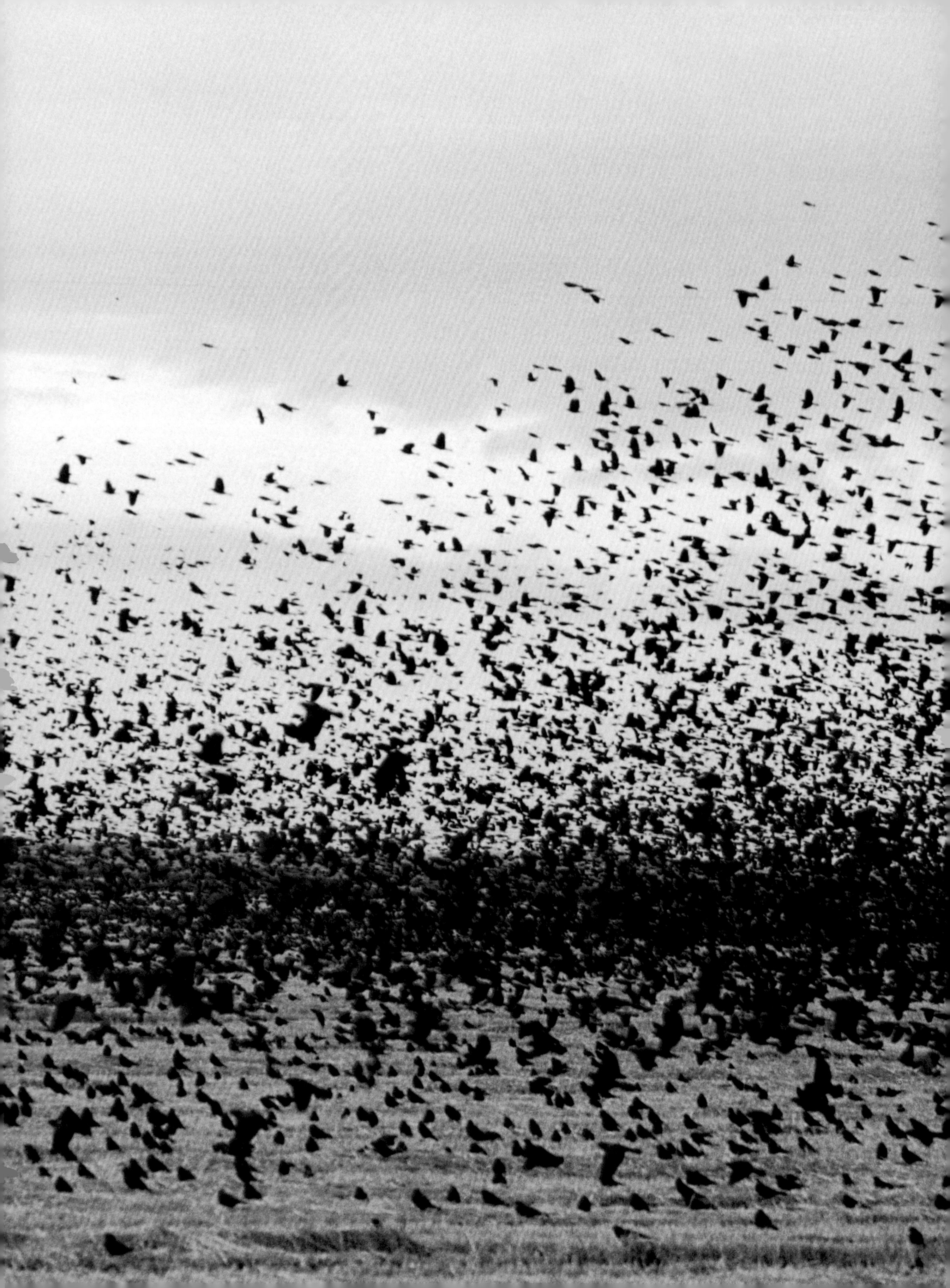

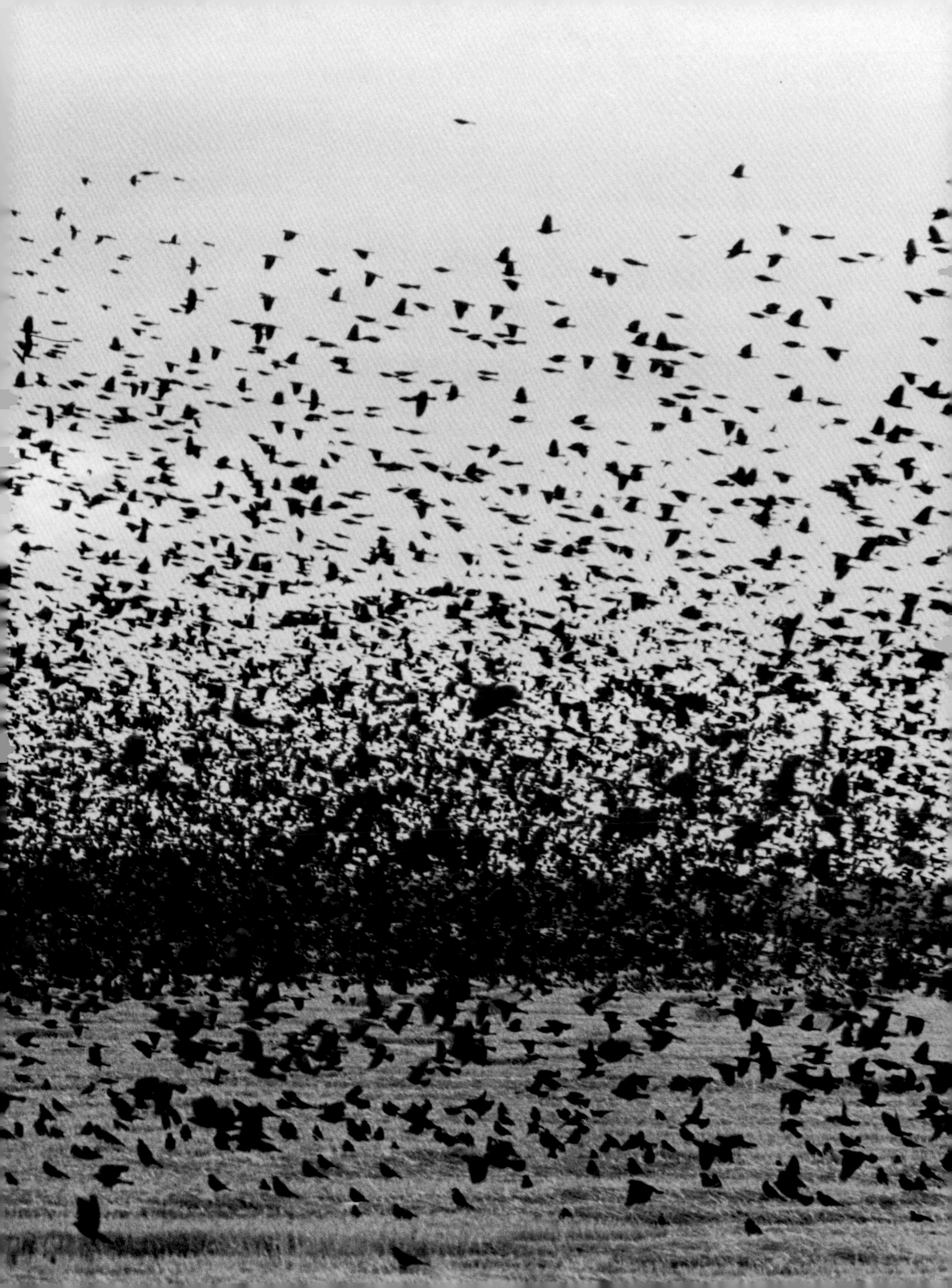

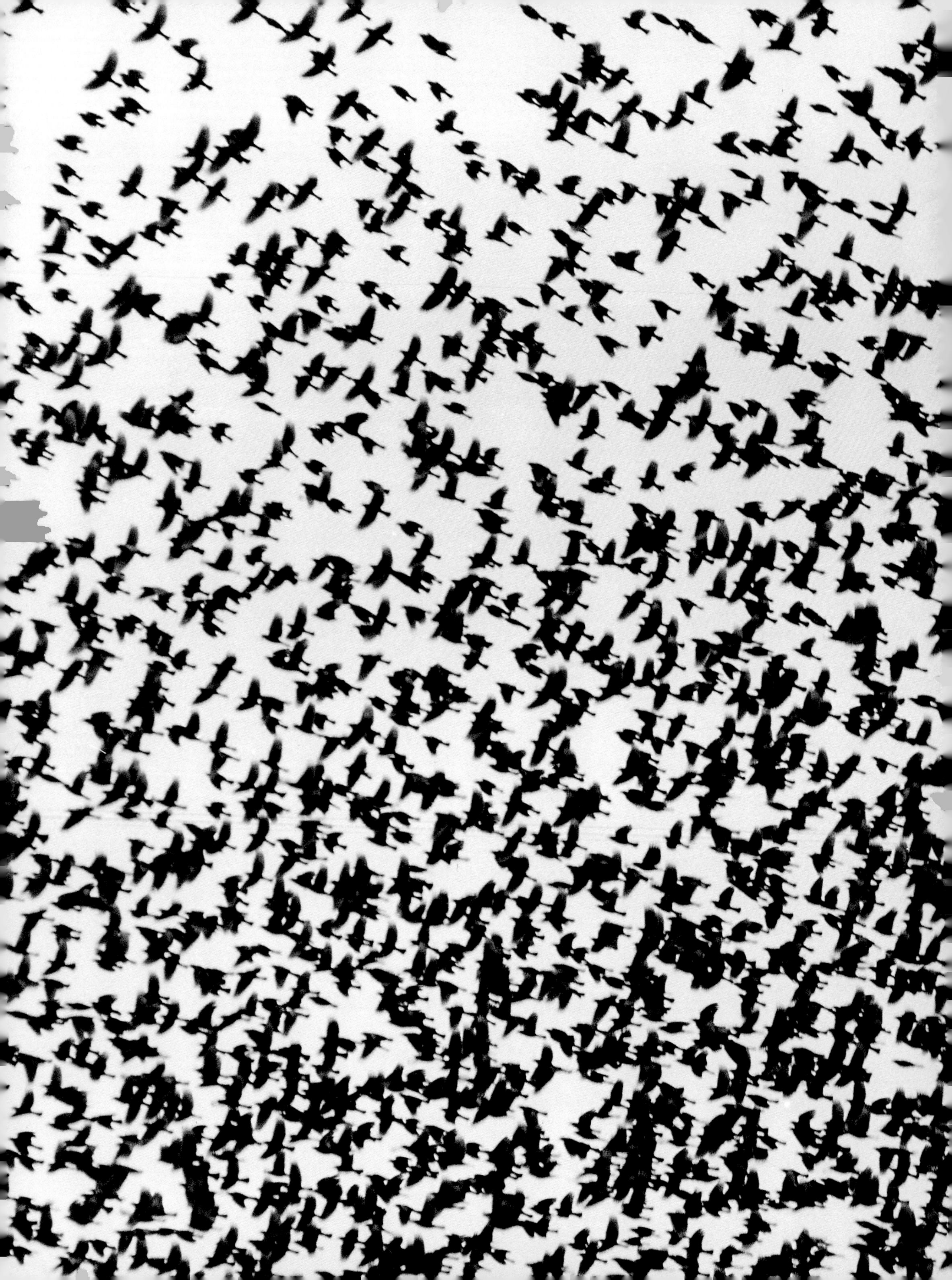

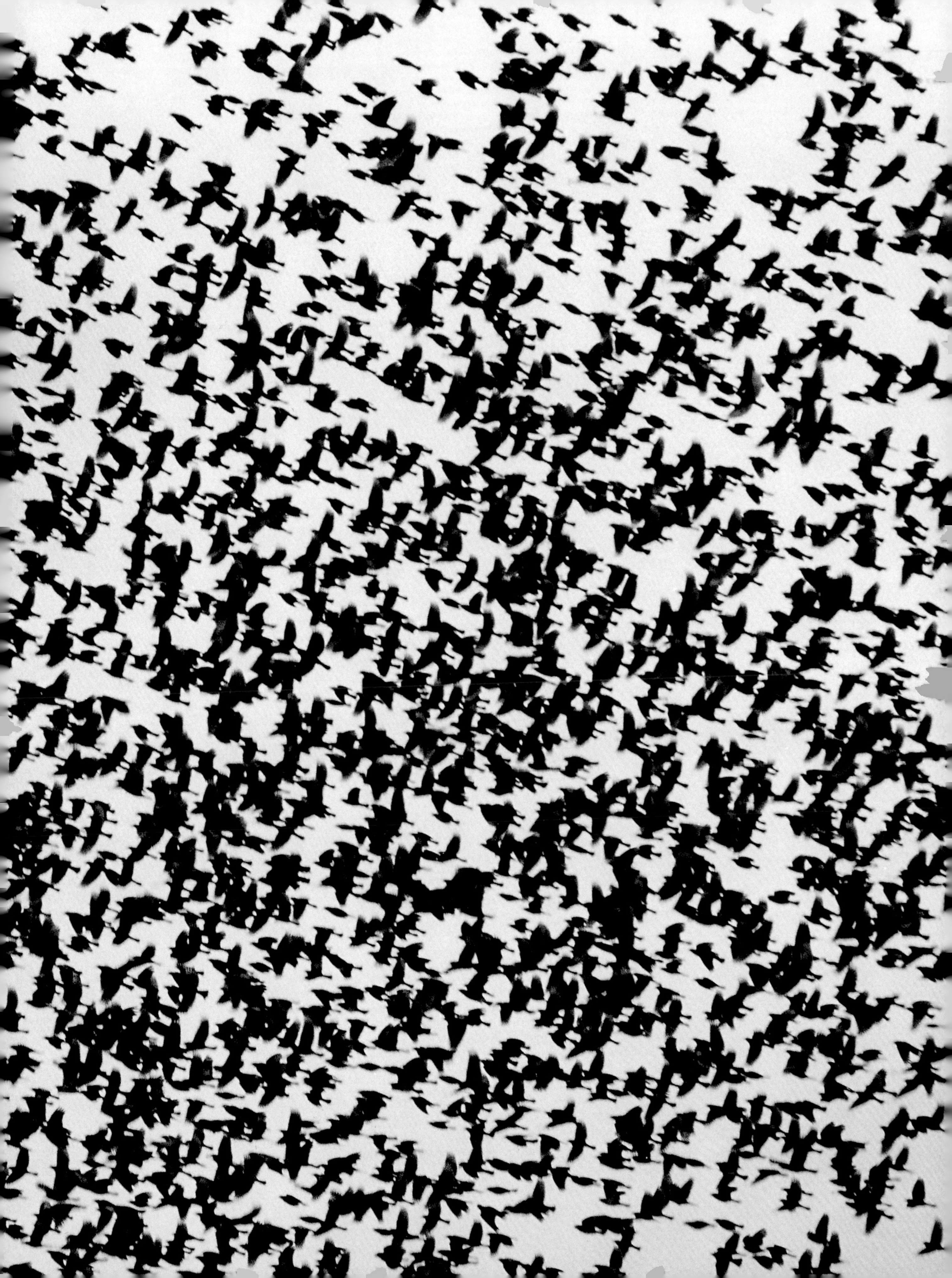

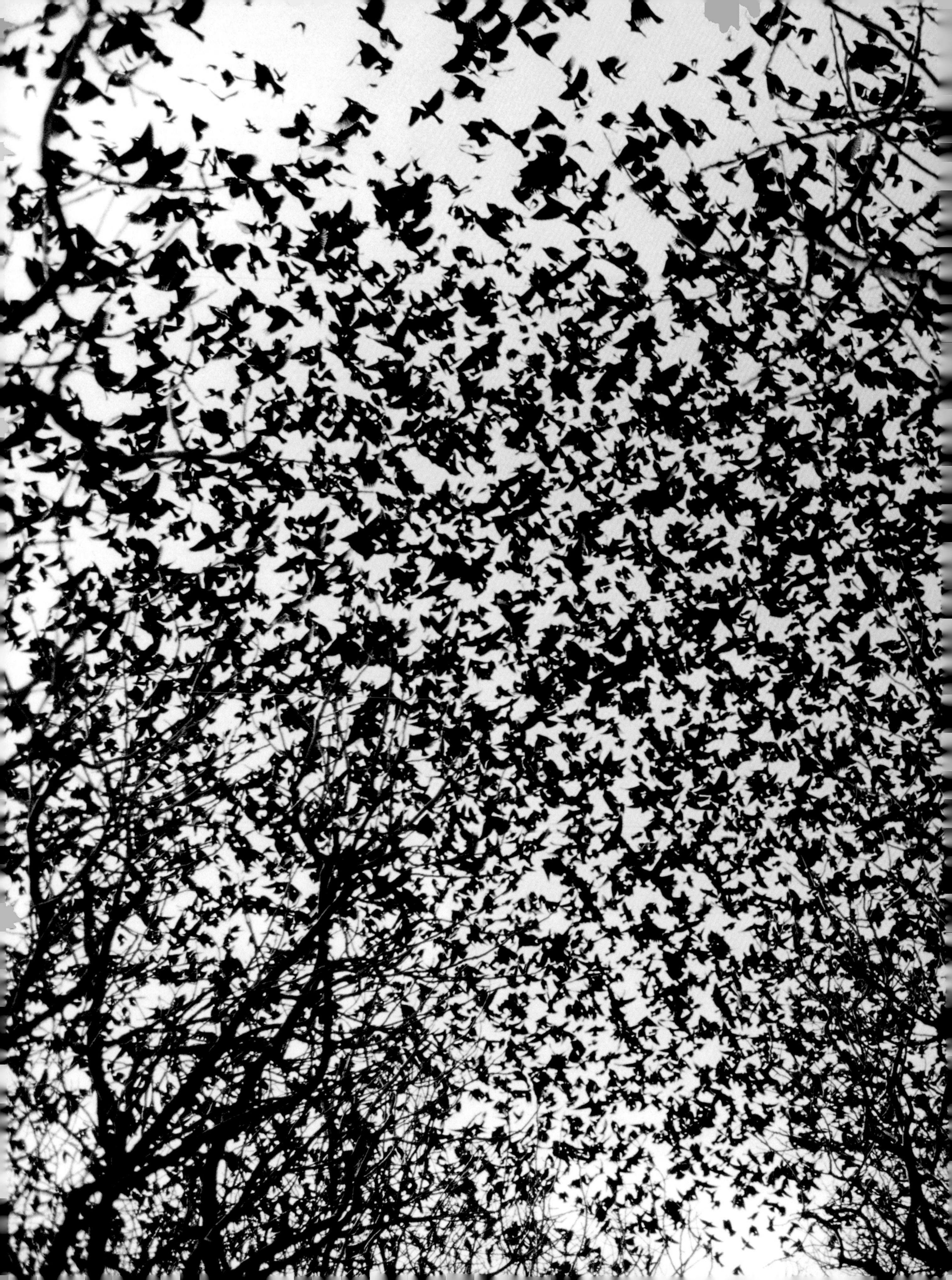

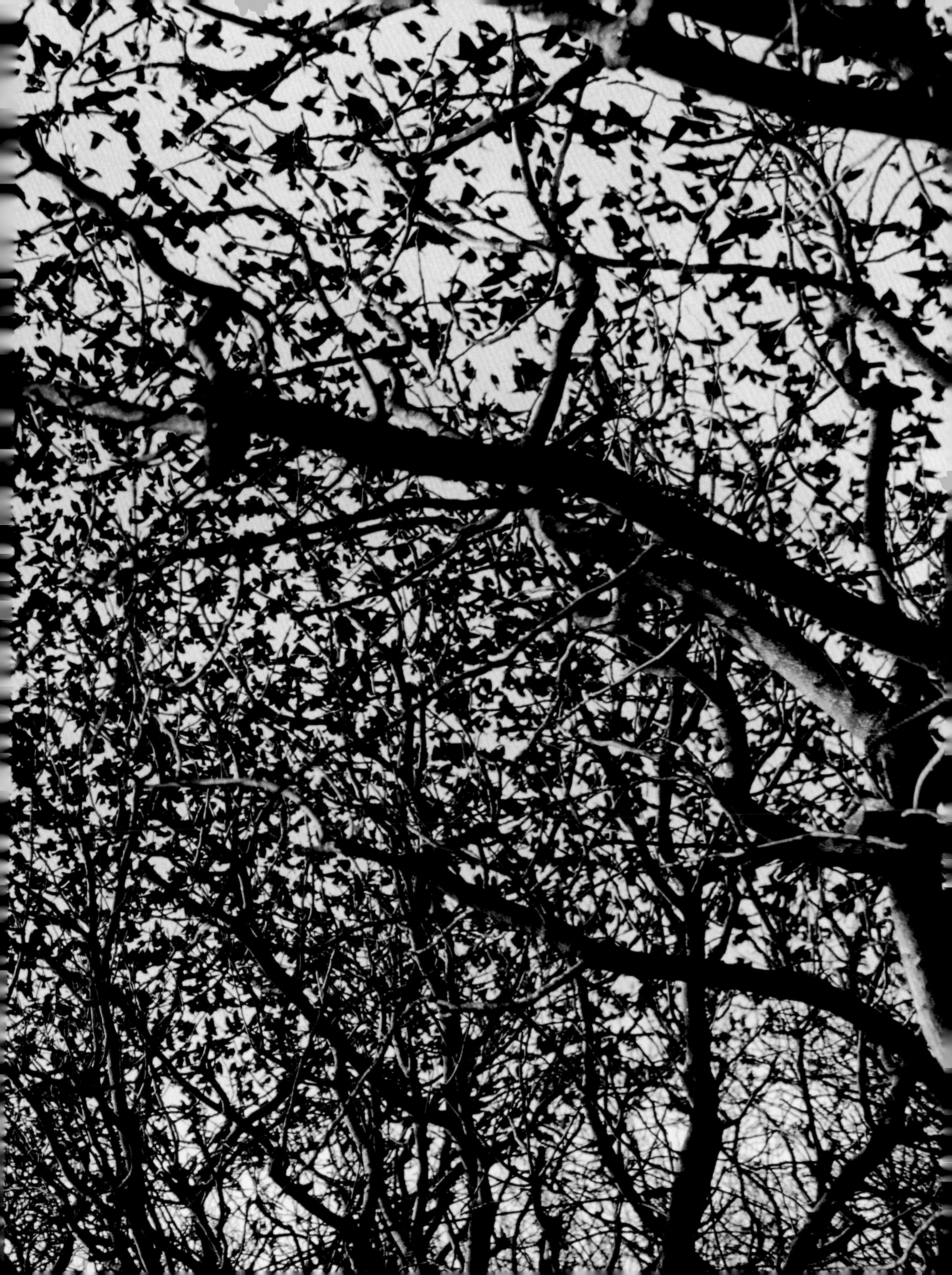

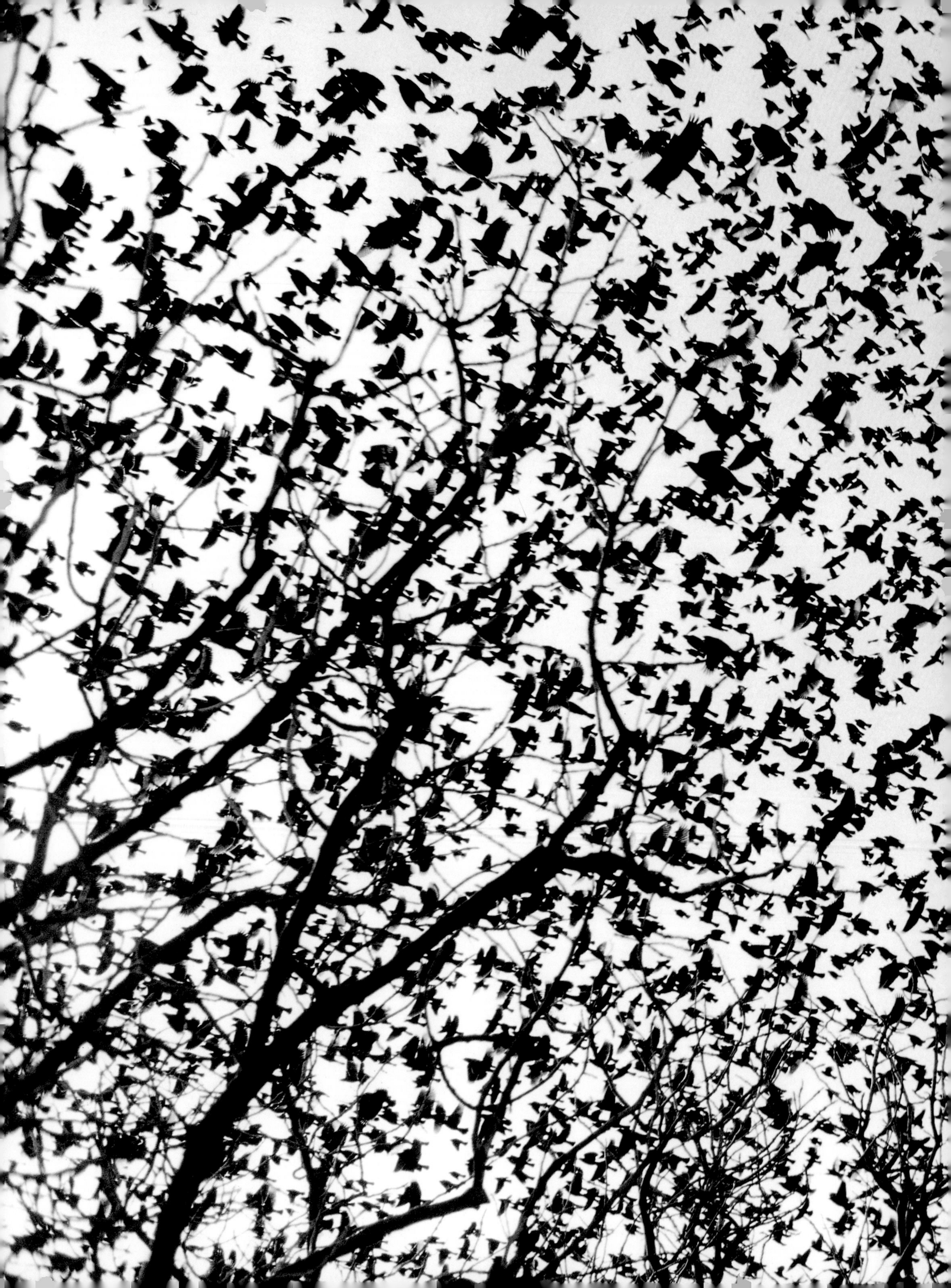

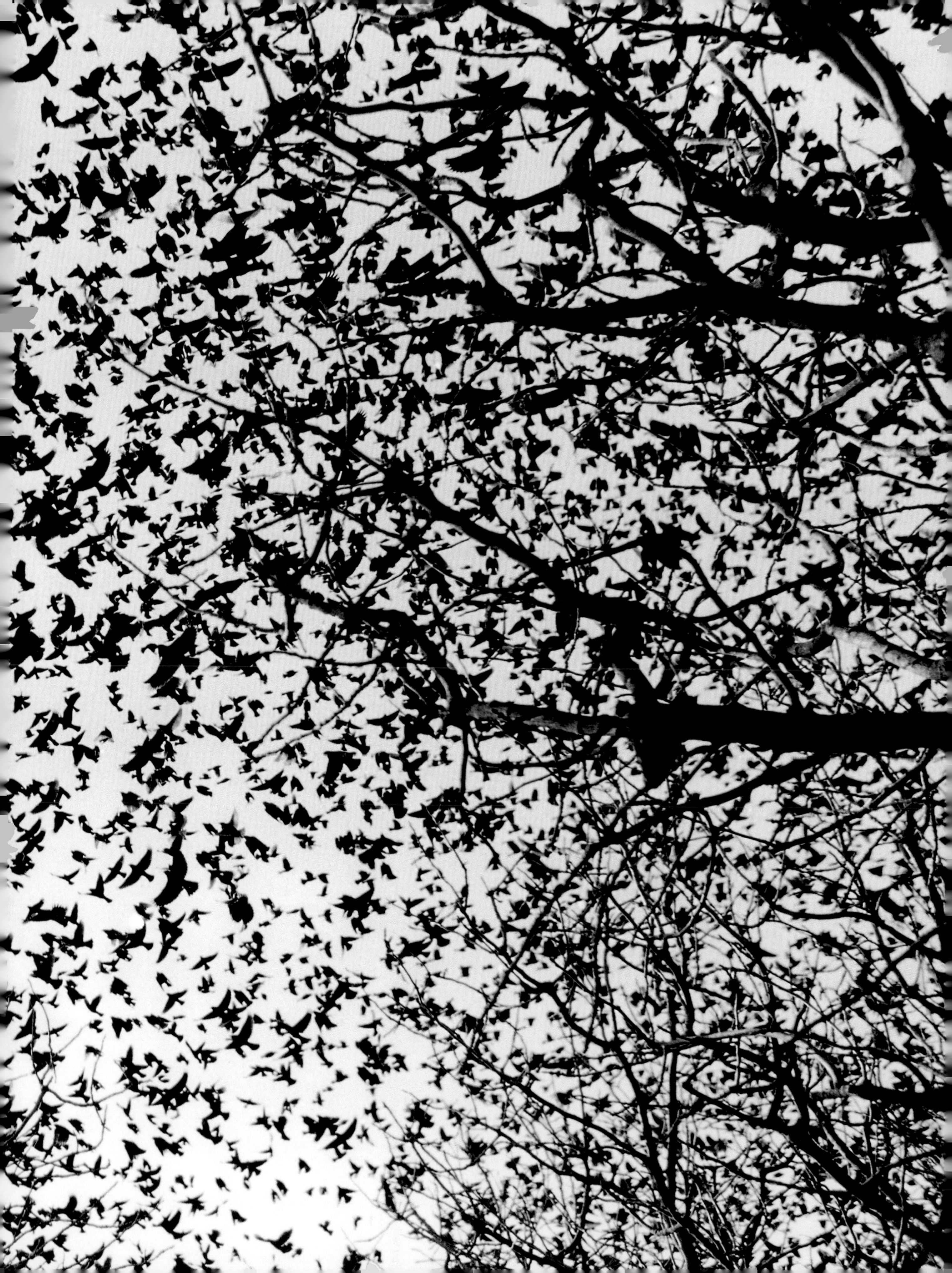

Lukas Felzmann, born in 1959 in Zurich, is an artist who lives and works in San Francisco. His work has been shown in Switzerland, Germany, the United States, Colombia, and Egypt. He teaches photography at Stanford University. His three previous books are *Landfall, Waters in Between,* and *Helix.*

Deborah M. Gordon, born in 1955 in Miami, Florida, is a Professor of Biology at Stanford University. She studies ant colony behavior and ecology in harvester ants in Arizona, argentine ants in California, and turtle ants in Mexico. She has written many articles and was awarded a Guggenheim fellowship. Her previous books are *Ants at Work* and *Ant Encounters.*

Gordon H. Orians, born in 1932 in Eau Claire, Wisconsin, is a behavioral ecologist and Professor Emeritus of Biology at the University of Washington. He currently writes on environmental aesthetics. His previous books are *Some Adaptations of Marsh-Nesting Blackbirds, Red-Winged Blackbirds: Decision-Making and Reproductive Success,* and *Blackbirds of the Americas.*

Peter Pfrunder, born in 1959 in Singapore, studied German and European literature in Zurich, Montpellier, and Berlin. He has worked as a journalist, writer, and curator and has authored many books on photography and photographers. He is currently Director of the Swiss Foundation of Photography/Fotostiftung Schweiz in Winterthur, Switzerland.

Wallace Stevens, born in 1879 in Reading, Pennsylvania, is regarded as an influential American poet. During his life he published nine books of poetry. *Harmonium,* which appeared in 1923, was his first book and included "Thirteen Ways of Looking at a Blackbird." In 1955, the year he died, he was awarded the Pulitzer Prize for Poetry and won his second National Book Award.

for Dora

Thanks to:
Gordon Orians for a lifetime study of blackbirds and beyond. Deborah Gordon for the patience with ants and the insightful text. Peter Pfrunder for his dedication to photography and for contributing an essay. Lars Müller for the continued collaboration and for believing in books. John Berger for looking through the manuscript and the passion and curiosity concerning all life. Larry Sultan for early inspiration as a teacher. George Steinmetz for giving me the lens I used for many of these pictures. Andre Salvage for balance, movement, and power through San Soo. The Fotostiftung Schweiz in Winterthur for expanding my archives there with a complete set of *Swarm*. All my colleagues at the Stanford Department of Art & Art History, especially Joel Leivick for his humor and Craig Weiss for logistical support. The Swiss Arts Council Pro Helvetia and the McNamara Fund at Stanford University for support. My family, Rhonda and Asa, for being with me.

Lukas Felzmann
Swarm

Design: Lukas Felzmann with Lars Müller
Copyediting and proofreading: Danko Szabó
Lithography, printing and binding: EBS, Verona, Italy
Paper: Tatami White 150 g/m²

Translation (text by Peter Pfrunder): Ben Fergusson
Quote by Novalis: *The Novices of Sais,* New York, 2005

Lars Müller Publishers
Baden, Switzerland
www.lars-muller-publishers.com

www.lukasfelzmann.com

ISBN 978-3-03778-241-5

Printed in Italy

With kind support of the Friends of the Fotostiftung Schweiz, Winterthur, and Pro Helvetia, Schweizer Kulturstiftung.

Published by Lars Müller Publishers:

Lukas Felzmann
Waters in Between, 2009
ISBN 978-3-03778-138-8

Lukas Felzmann
Landfall, 2004
ISBN 978-3-907078-92-1

When the blackbird flew out of sight,
It marked the edge
Of one of many circles.
W.S.